KV-038-899

Ecotoxicology of Organic Contaminants

Eros Bacci

Library of Congress Cataloging-in-Publication Data

Bacci, Eros.
Ecotoxicology of organic contaminants / Eros Bacci.
p. cm.
Includes bibliographical references and index.
ISBN 1-56670-022-1 (acid-free paper)
1. Environmental risk assessment. 2. Pollution—Environmental aspects. 3. Organic water contaminants—Environmental aspects. 4. Environmental chemistry. I. Title.
GE145.B33 1993
628.5′2—dc20 93-8849
CIP

Direct all inquiries to CRC Press, Inc., 2000 Corporate Blvd., N.W., Boca Raton, Florida 33431.

Lewis Publishers is an imprint of CRC Press

International Standard Book Number 1-56670-022-1
Library of Congress Card Number 93-8849
Printed in the United States of America 1 2 3 4 5 6 7 8 9 0
Printed on acid-free paper

For Faro and Clementina

Preface

The human impact on natural systems is becoming heavier and heavier in both technologically advanced and developing countries. Some of the resulting pollution problems are of global scale, giving rise to measurable modifications in the air, water, soils, and living organisms, the long-term effects of which are unknown.

There is an increasing need to improve present knowledge on key aspects of the environmental distribution and fate of contaminants and of the related effects on natural ecosystems. This book shows some of the feasible approaches to the assessment of the environmental fate of organic chemical contaminants, together with available instruments to evaluate toxicity both at the single-species and at the biological community level. The ultimate goal is to define value and limitations of present hazard and risk assessment procedures.

The book does not represent a comprehensive review of the abundant literature on the topics of environmental chemistry and environmental toxicology. In it are simply indicated the views of the author in selecting strategies to be applicable not only to a limited number of chemicals but, in principle, to all organic contaminants — a sort of multiresidual approach, where the physicochemical and toxicological properties are considered, rather than the compounds themselves — in the belief that such an approach can greatly reduce the complexity of the problem. The problem is actually tremendously complex: we are trying to understand the environmental behavior of *mixtures* of chemicals, in *mixtures* of environmental compartments (e.g., air, water, soil, biota), interacting with *mixtures* of organisms and populations (i.e., biological communities).

Current knowledge is still very limited. However, in the last 20 years, great improvements have been made in tools for predicting the environmental fate of contaminants, in environmental toxicology and in hazard assessment procedures. During my teaching of undergraduate and postgraduate applied ecology and ecotoxicology, I have found that young students (much more so than old colleagues) are greatly interested in these aspects, and pose questions that are not always easy to answer. In this book, the attempts to answer these questions are presented in order of growing complexity, in a manner that intends to inspire further questions and answers. According to the main components of ecotoxicology, the book is divided into three sections: assessment of environmental fate, environmental toxicology, and hazard and risk assessment.

In each section, rather than *the solution* to the various problems, possible ways to approach them are shown, trying to avoid oversimplifications, but at the same time avoiding excess of *details*, which are left to the various disciplines on which ecotoxicology is based. Errors, omissions, and oversights are the fault of the author, to whom the book is primarily addressed: he still has to learn how to apply an *ecotoxicological approach* in studying contamination. The main difficulty probably consists in reaching an acceptable equilibrium between the science of experimental measurement, observation, and data gathering, and the art of extrapolation simply based on the validity of a hypothesis. The final judgment is left to the readers!

Eros Bacci received his degree in Biological Sciences from the University of Siena (Siena, Italy), where he began studying environmental contamination problems in 1972 and where he is now professor of Applied Ecology and Ecotoxicology in the Department of Environmental Biology. Research interests concern the role of physicochemical properties in determining the environmental fate of contaminants and the applications of evaluative models. He has organized and conducted ecotoxicology research programs in Italy and other Mediterranean countries as well as a global survey on some conservative contaminants. Professor Bacci has published 99 articles, reports, and book chapters in scientific literature.

Contents

Section 1. *Approaches to the Assessment of the Environmental Fate of Chemical Contaminants*

Section 2. *Environmental Toxicology*

> Unlike the surface waters, which are sensitive to every gust of wind, which know day and night, respond to the pull of sun and moon, and change as the seasons change, the deep waters are a place where change comes slowly, if at all. Down beyond the reach of the sun's rays, there is no alternation of light and darkness. There is rather an endless night, as old as the sea itself.
>
> Rachel Carson

SECTION 1
Approaches to the Assessment of the Environmental Fate of Chemical Contaminants

1.1. INTRODUCTION

The first approaches to the study of environmental contamination were essentially directed toward preserving human health from the possible backlash of harmful substances released in the atmosphere, soils, or water bodies. Later, the idea that human health cannot be protected unless in conjunction with wildlife protection, led to a definition of a new branch in environmental sciences, *ecotoxicology*. The term was introduced by Truhaut in 1969 (see Truhaut, 1975) who defined ecotoxicology as a branch of toxicology dealing with direct or indirect effects of natural substances and artificial pollutants on living organisms. This implies that environmental chemistry is also involved, providing the instruments to understand the environmental distribution and transformations of chemicals (Koeman, 1983). Ecology is obviously implied in the evaluation of effects on the ecosystems, which substantially differ from the effects on individual organisms or populations: for instance, the death of an individual or the disappearance of a population are dramatic for whoever dies or disappears, but may have no significant influence on the functions of an ecosystem. On the other hand, subtle sublethal effects on certain species may produce significant ecological modifications.

The goal of ecotoxicology consists of producing criteria for the prevention or minimization of contamination. Due to the rapid evolution of technology, ecotoxicology should also apply predictive instruments, able to produce criteria even for hypothetical chemicals or potential contaminants, before their release into the natural system. According to John Cairns, Jr. (1980), ecotoxicological *criteria* and *standards*, often regarded as synonymous, should be viewed as two different concepts: criteria are the product of scientific investigation, while standards are the product of laws and regulations. Standards are derived from the political combination of different criteria (ecotoxicological, economical, technological, social, etc.). The indication of a criterion is a matter for science, the adoption of a standard is a matter for governmental authorities.

The number of chemicals in common use, apart from pesticides, pharmaceuticals, and food additives, is estimated at 50,000; all included, the total is higher than 60,000 compounds (Cairns, 1980), with between 500 and 1000 new ones added to the list each year (Blum and Speece, 1990). A great number of synthetic organic substances, such as detergents, solvents, plastics, and antifouling agents, are continuously introduced into the environment in large quantities. In the more industrialized countries, these quantities are of the same order of magnitude of primary production by plants (300 g organic matter per square meter per year; Stumm et al., 1983).

The aim of this book is to outline the instruments available for the study of the environmental fate and effects of chemical contaminants, with a particular attention to the organic chemicals. The main objective remains to contribute to answer the fundamental question of *"whether we can sustain the natural systems which permit us to persist"* (Hope, 1993).

1.1.1. Contamination and Pollution

As far as environmental contamination and pollution are concerned, several definitions have been produced. The following definition of air pollution appeared in 1967 (*cf.* Trouhaut, 1975) in a Council of Europe report:

> "Pollution of the air occurs when there is present in the air a foreign substance or an important variation in the proportions of its constituents capable of causing a harmful effect or of causing discomfort, bearing in mind the extent of scientific knowledge at the time."

A definition of marine pollution was provided by the IMO/FAO/UNESCO/WMO/WHO/IAEA/UNEP Joint Group of Experts on the Scientific Aspects of Marine Pollution (GESAMP):

> "Pollution of the marine environment means the introduction by man, directly or indirectly, of substances or energy into the marine environment (including estuaries) which results in such deleterious effects as harm to living resources, hazard to human health, hindrance to marine activities including fishing, impairment of quality for use of sea water and reduction of amenities."

These definitions, directed to a selected environmental compartment (air and the marine environment), appear today quite inappropriate due to the multimedia nature of the environment.

In ecology, *conditions* refer to abiotic environmental factors which are not depleted by biological activity; *resources* are normally exploited by living organisms reducing their availability to other organisms, but not altering their quality. Contamination is a change in one or more of the environmental conditions such as temperature, acidity, transparency caused by an agent. Contamination is also a change induced by man in the availability of resources (e.g., eutrophication) or the alteration of their properties (pesticide residues in fish). The exploitation of resources by living organisms does not constitute contamination.

In the light of the above concepts, *environmental contamination can be defined as a human action able to modify properties of environmental conditions or the availability and quality of resources over a given space range and time interval.* Environmental contamination does not necessarily imply measurable damage to living organisms.

Organisms with similar characteristics constitute a species, or the realization of a particular genetic potential in a given time interval; a group of organisms of the same species living in a particular area is a population (e.g., trout in a lake, human beings in a country or in the world, white oaks in a forest). Different populations living and interacting in a particular site constitute a biological community (e.g., a forest with higher plants, animals, and microorganisms).

In the case of an impairment of a biological system (organism, population, or community), *environmental pollution* is occurring (Moriarty, 1983; Royal Commission, 1984):

> *Environmental pollution occurs when contamination produces measurable damage to single organisms, populations, or biological communities.*

Damages and their measurement will be discussed in Section 2.

Environmental contamination and pollution phenomena can be more or less limited in space and time, from local to global, from very short to very long. Pollutions of short duration and limited in space can be regarded as *ecological disasters,* such as a storm or a forest fire, to which natural systems can react, rapidly recovering their previous structure and functions. Human activities, unfortunately, often imply extensive and persistent environmental contamination and pollution. At present, it is still very difficult to predict the long-term effects of some pollutants *(ecological catastrophes?).*

It is essential to realize that contamination, *per se,* is an environmental impairment. Consequently, all actions should be directed, whenever possible, to prevent or at least to minimize contamination.

1.1.2. Perception

World human population is increasing exponentially. It doubled from the year 0 to 1500, reaching 600 million people. A second doubling occurred in 1800, reaching 1.2 billion, followed by another doubling in 1950 (2.5 billion). In 1980, the world population was roughly 5 billion. This clearly indicates that the growth is accelerating, mainly sustained by agricultural and industrial technology. At present, the growth rate is 1.7% per year, corresponding to a doubling time of 41 years (Tyler Miller, 1989). Considering the total surface of our planet (510,000,000 km^2), present average population density is around 10 inhabitants per square kilometer, or 10 inhabitants per 100 ha. Since only part of these 100 ha is land, and only part of this land is available and compatible with human settlement, the actual average world population density is in the order of 1 person/ha. The high population density constitutes the background for the crisis in the relations between Man and the Environment.

The increase in energy production (mainly based on external sources), the use of an exponentially increasing number of vehicles for transportation, modified agricultural practices, and the large-scale introduction of synthetic chemicals, previously unknown to biota *(xenobiotics)*, are additional factors of crisis. Moreover, human pressure on the environment is not equally distributed: from one hand, 23% of the world's population in the more developed countries is using 80% of the world's processed energy and mineral resources; on the other hand, the less developed countries are rapidly increasing their environmental loads, often with obsolete technologies, in an attempt to rise their economic level. Urban areas are rapidly growing in both types of countries, adding another imbalance factor. One of the more evident critical points is the growth of the urban water demand, which is so rapidly increasing that even the more technologically advanced countries are in trouble: the water demand in southern California, for instance, exceeds locally available and imported supplies (the latter are 125 out of the 200 gallons daily used per capita) and, by 2010, a supply shortfall of more than 30% is expected (Dziegielewski and Baumann, 1992).

A comparison between the "natural man" and the "industrial man" annual load of nitrogen, sulphur, and carbon dioxide in different countries is reported in Table 1.1. It can be seen that even the per capita annual load of natural chemicals such as nitrogen, sulphur, and carbon dioxide is greatly increased by human technological activities, with some serious risk of modifying natural cycles.

Perception of the problems related to environmental contamination began during the 1950s and became popular early in the 1960s, after the publication of the best seller *Silent Spring* by Rachel Carson (1962), one of the best examples of mixing art and science, with no price for anyone of them. The possibility of seeing our planet from outside as a small and vulnerable space vehicle probably contributed to the perception of the limited dilution and recovery potential of the system and the significance of contamination problems. The physical limitations of the planet led to a shift from the past idea of possible compartmentalization within nations or within different environmental matrices (water, air, soil) giving way to a new vision of the planet as an *unit* in which problems originating in one site or sector may significantly influence other sites or sectors, even at a great distance. In the 1960s, public concern was still far from the concept of *sustainable development,* or development able to solve present problems without prejudicing the possibility of future generations to do the same. However, it was very clear that human activity could modify structural and functional properties of the environment. The discovery of the effects of methyl mercury wastes on people eating contaminated fish and shellfish in the Minamata Bay in Japan, and the presence

Table 1.1. Annual Environmental Load of Nitrogen, Sulphur, and Carbon Dioxide by *Natural* and *Industrial* Man in Some Countries

				Natural man (kg/(ha·yr))			Industrial man (kg/(ha·yr))		
Country	Population (1000s)	Area (1000 km²)	m² per inhab.	N	S	CO_2	N	S	CO_2
Netherlands	14220	41	2883	13.9	1.7	1040	73.9	34.4	31200
U.K.	56763	241	4246	9.4	1.1	707	39.6	80.3	22800
Italy	57140	301	5268	7.6	0.9	570	27.8	39.4	11900
France	53788	547	10170	3.9	0.5	295	22.1	13.9	6900
Turkey	45218	781	17272	2.3	0.3	174	8.0	2.3	1400
Sweden	8320	450	54087	0.7	0.1	55	3.2	2.6	1300

Note: The data on natural man is based on the standard production *pro-capite* and per year of 370 l of urine containing 3.4 kg N and 0.5 kg S, 12 kg of feces (dry weight) containing 0.6 kg N, and 300 kg of CO_2.

Partially reproduced from Blank et al. (1992). With permission.

of the insecticide DDT and related compounds in declining animal populations living in site where it was never used, contributed to this awareness.

Awareness of environmental contamination was made possible by the rapid development of new analytical techniques, such as atomic absorption spectrophotometry and gas chromatography, able to detect traces of contaminants in various environmental matrices. During the 1960s and the first half of the 1970s, an increasing number of scientific papers was published on the detection of microcontaminants in various environments. The focus was on *where* and *how much*, essentially based on a retrospective approach.

1.1.3. The Retrospective Approach

The retrospective approach consists of the study by *monitoring* of the environmental distribution of contaminants already in the environment. Monitoring can be defined as the field measurement of the levels of one or more contaminants *(chemical monitoring)* or of the biological effects of a pollutant *(biological monitoring)*. Monitoring includes *surveys* (a single set of data on a spatial grid and sporadic time basis) or regular *surveillance* (based on temporal and spatial grids) with the focus on collecting information on environmental levels and their variations in space (survey) or in space and time (surveillance; Wells and Côté, 1988).

The results obtained were extremely useful for seeing which chemicals were generating contamination or pollution, where this was occurring, and any eventual impairment to wildlife and man. Since the 1960s, this has led to a great number of environmental laws and regulations, particularly in the more technologically advanced countries. Limits (often expressed as *concentrations*) for effluent control in water and in air were introduced and *black, grey,* and *white lists* of hazardous substances produced (van Esch, 1978). The use of very dangerous substances was severely restricted early in the 1970s in North America and Europe.

The historical merits of the retrospective approach are unquestionable, having provided baseline information on the main pollutants and on the location of polluted sites, and even revealing some global contaminants, such as the famous DDT. In many cases, the adverse effects in polluted environments were effectively mitigated by means of recovery actions. This approach also enabled the identification of unknown or new contaminants: polychlorinated biphenyls (PCBs) were discovered, by chance, by Jensen in 1964 as *spurious peaks* in extracts of muscle of white-tailed eagles, and identified 2 years later (Jensen, 1972).

The major limitation of this approach, however, lies in its poor interpretation potential. With few exceptions, most old scientific papers provide tables of data and descriptions of phenomena rather than interpretations. The focus is, for instance, on the levels of mercury and lead in fish muscle, without any clear connection with the possible contamination source, environmental pathways, and biological meaning of the found levels. The key point in interpreting results was whether they were *"higher, equal, or lower than background or reference levels."*

In the worst case, the contaminant to be studied was decided by the availability of instruments and analytical techniques. Global contaminants are still stimulating the *discovery* of new organisms to be used as *indicators* of contamination! The use of organisms as bioindicators of contamination has sometimes been carried out without testing the accuracy of the approach; as stated by Phillips and Segar (1986), *"there is no justification for simply assuming that any new untested species can act as an efficient bio-indicator."*

1.1.4. Less *"Where?"*, More *"How?"* and *"Why?"*

In mid 1980s, an important international scientific journal, the *Marine Pollution Bulletin,* published a viewpoint by David A. Wright (1985) entitled: *Less "Where?", More "How?" and "Why?"* in which the need to understand more about the environmental fate of contaminants was stressed, and studies restricted to reporting tables of concentrations discouraged: *"we can no longer simply go out and measure things."* The paper suggested that the retrospective approach had exhausted its possibilities and needed to be replaced by new predictive instruments. At that time, these were still quite crude and probably need, even today, to be greatly improved. However, most of the monitoring studies had shown their weak point: designed to answer questions such as *"what could we do to reduce present pollution in that area?"* produced, sometimes after years of study, the following answer: *"more monitoring!"*

Monitoring can of course be very useful for measuring the need for, and then the effects of, a regulation, but it needs always to be well directed and planned to avoid misuse of scientific and economic resources. Field studies, when correctly planned, are crucial to point out essential processes determining the environmental fate of contaminants, particularly when combined with conceptual models and selected laboratory experiments. The lesson of the discovery of PCBs, mentioned above, indicates the need for improving predictive instruments in order to minimize the future possibility of similar discoveries. It is now well known that these compounds, because of their physicochemical properties (hydrophobicity, sufficient mobility and scarce chemical reactivity), behave very similarly to the most famous DDT (for this reason, they are called *DDT-like compounds*). In fact, the pollution potential of different substances mainly depends on their physicochemical properties (Hutzinger et al., 1978).

1.1.5. The Need for Prediction

The National Environmental Policy Act, NEPA, was approved in the U.S. in 1969. In force in 1970, this federal law (Public Law 91-190 Jan. 1st, 1970) introduced the Environmental Impact Assessment (EIA) procedure (Munn, 1975), requiring all federal agencies to prepare an Environmental Impact Statement *prior* to taking any major action or adopting legislation that would significantly affect the natural system.

Less formal impact assessment was carried out before NEPA. For example, in England in 1548, a commission was set up to examine the environmental impact of the Wealdon iron mills and furnaces in Kent and Sussex. As reported by McDonnell (1992), this commission was required to evaluate the effects of the proliferation of these plants on the economy of the region and the impact of wood consumption on forests.

The NEPA approach was accepted in principle by the UN Conference on the Human Environment held in Stockholm in 1972. In the early 1970s and 1980s, EIA procedures were adopted in several technologically advanced countries (e.g., Japan, Australia, Canada, United Kingdom, and France; Munn, 1975) and by the European Economic Community (EEC).

With NEPA, the scientific community was requested to provide predictions which, at that time (and still today), were considerably difficult, given the limitations of available knowledge.

A few years later, another U.S. law came into force: the Toxic Substances Control Act (ToSCA; Public Law 94-469-Oct. 11, 1976-90 STAT 2003). The law was aimed at protecting man and the environment, considered jointly, from hazardous substances. Clearance from the U.S. Environmental Protection Agency (EPA) is mandatory in order to manufacture a new chemical or process a chemical substance for a use not previously registered. The main feature of the ToSCA is to establish a procedure for estimating the hazard *before* marketing. The hazard of the chemical is considered from its extraction to its manufacture, distribution, processing, use, and disposal. The procedure is based on principal physicochemical properties, toxicological and ecotoxicological studies, and environmental chemistry investigations. The adoption of the ToSCA significantly contributed to the progress of predictive approaches and to the definition of strategies.

The EEC adopted a similar regulation in 1979 (Council Directive 79/831; Off. Jour. Eur. Comm. L 259/Oct. 15, 1979). In this document, a revision of "old" chemicals, as in the ToSCA, is not included; however, the kind of information requested before marketing a new chemical is quite similar to that of the ToSCA. In more recent years, a continuous refining of environmental protection criteria has been produced, particularly for pesticide hazard and risk assessment (Klein et al., 1993).

1.2. PREDICTIVE APPROACHES AND MODELS

The need for prediction forced environmental chemists and ecotoxicologists to produce conceptual models for the analysis of the various problems, instead of collecting data from chemical analysis (the *grid-based approach*).

Models are tools for the analysis of the role of the properties of environmental systems and of properties of chemicals in determining their environmental fate. Models are also tools for synthesis, providing overviews of problems and key aspects regulating the environmental fate of contaminants. By definition, models can never entirely represent a real system, but they can be used to get *essential* information on the system. Models can be either laboratory models or physicomathematical models.

1.2.1. Laboratory Models

Laboratory models contain selected components of a real system with the aim of experimentally observing the behavior of substances (reactivity, mobility) under simplified conditions.

More complex models can be divided into microcosm, mesocosm, and macrocosm systems, according to their volume and their realism; in these, a more or less simplified natural environment is reproduced on a reduced scale. Often, these models are still too complex, transferring to the laboratory too many of the variables of the field. Metcalf and co-workers (1971) introduced a simplified model laboratory ecosystem, which was subsequently applied to simulate the fate of more than 100 chemicals (Figure 1.1.)

Very simple laboratory models have been successfully applied to point out some key aspects of the environmental behavior of contaminants. As an example, Figure 1.2 shows one of the greenhouses used by Bacci and Gaggi (1985) to answer the question: "What is the significance of root translocation in generating levels of polychlorinated biphenyls found in plant foliage?" In the model, the soil was a sand and plants were broad beans, beans, tomatoes, and cucumbers. Plants grown in "clean" and PCB-fortified soil were kept in the same air, contaminated by PCB vapors coming from the fortified soil. The answer was clear: levels in foliage were not dependent on contamination of the soils where the plants grew.

Of course, this very simple model does not permit the exclusion of some root translocation, but shows that vapor movements from the soil play a major role in the levels found in foliage. The exclusion of root translocation may be derived from some simple theoretical considerations, based on liquid chromatographic principles: hydrophobic substances are characterized by a very scarce mobility in water (McCrady et al., 1987; Ryan et al., 1988). Experimental support for this theory is available in the literature. For instance, Schroll and Scheunert (1992)

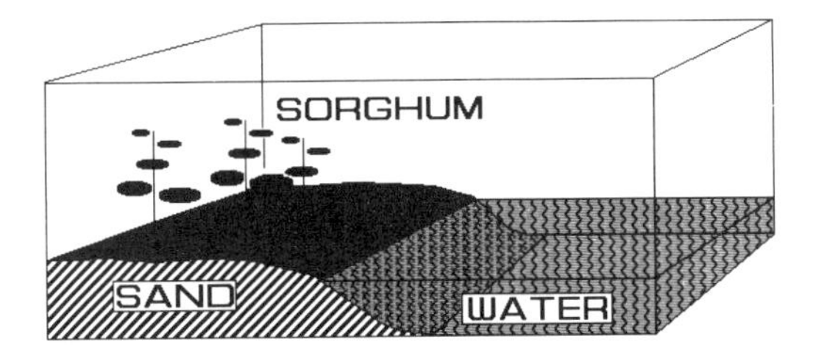

Figure 1.1. A simple model ecosystem. (Modified from Metcalf et al., 1971.)

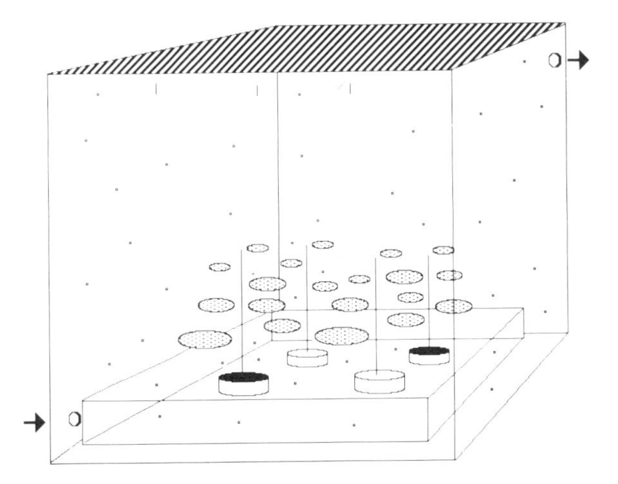

Figure 1.2. A simple laboratory model for studying root translocation vs vapor uptake in plant leaves. (Modified from Bacci and Gaggi, 1985.)

tested the mobility of a nonpolar chemical (hexachlorobenzene) in various plant species: all residues found in roots were due to root uptake directly from the soil and all residues in shoots were due to foliar uptake from the contaminated air.

Simple models may be used to evaluate the relative significance of the reactivity and mobility of a pesticide in water or soil. The same apparatus in Figure 1.2, without plants, can be applied as an open system for measuring degradation of chemicals in soil. If the mass transport by air and water is taken into account, the *degradation rate* can be calculated from the detection of the quantities remaining in the soil at different times. In this case, the overall degradation rate is obtained, without any indication of the possible mechanisms of chemical reaction (hydrolysis, oxidation, photolysis, or biodegradation). When mobility of the chemical is reduced to low or negligible levels, measurements of chemical reactivity are directly obtained. This is less feasible in field conditions where the *disappearance* of chemicals is currently evaluated. *Disappearance is the result of chemical transformation and displacement.*

A highly water-soluble pesticide can quickly leave the treated soil, rapidly disappearing from this environmental compartment and displacing the problem to the water.

As it can be seen from this very simple discussion, laboratory models need to be coupled with physicomathematical models which provide the theoretical background to support experimental planning, data processing, and interpretation of results.

1.2.2. Physicomathematical Models

These models are based on physical laws, their formal expressions (working equations) and other mathematical elaborations used to calculate results. Skill in environmental modeling does not consist in solving more or less complex algorithms, but in the selection of the best set of assumptions.

Models consist of five components (Jørgensen, 1990):

- *forcing functions:* these are also called control functions or external variables; typical examples are temperature of the system (influencing process rates), in- and out-flows or, for closed systems, loads.
- *state variables:* describe the state of the system (e.g., selected environmental compartments).
- *mathematical equations:* indicate the relationships between the forcing functions and the state variables.
- *parameters:* coefficients of the mathematical equations; they can be considered constant for a specific simulation.
- *universal constants:* these are characterized by fixed numerical values and units, such as the gas constant or molar masses.

In principle, before the formal mathematical expression of the model, a block diagram is prepared illustrating all the components and relationships. A simple example illustrating the uptake and release of a chemical in fish is shown in Figure 1.3.

If first-order kinetics are applicable, the variation of the concentration of the chemical in the fish, C_F, mol/m^3, is expressed by the differential equation:

$$dC_F / dt = k_1 C_W - k_2 C_F \qquad (1.1)$$

where C_W is the concentration of the chemical in the surrounding water, mol/m^3, k_1 the input rate constant (1/h)*, and k_2 the clearance rate constant (1/h). When C_W = 0, Equation 1.1 becomes:

$$dC_F / dt = -k_2 C_F \qquad (1.2)$$

which is the typical first-order decay function (as for radionuclides) where the elimination rate is directly proportional to the concentration in the fish. The

* When C_F and C_W are not expressed in the same dimensions, k_1 may have different expressions. For instance, with C_F in mass/mass and C_W in mass/vol, k_1 is $\frac{\text{vol}}{\text{mass}} \times \frac{1}{\text{t}} \left(\text{e.g.,} \frac{\text{L}}{\text{kg}} \times \frac{1}{\text{h}}\right)$.

Figure 1.3. A block diagram for chemical uptake and release in fish.

elimination rate constant k_2 is always positive: the sign "–" is relevant to the direction of the chemical flux (away from the fish). Equations 1.1 and 1.2 are typical working equations.

If C_W is constant, Equation 1.1 may be solved as follows:

$$C_{F(t)} = k_1 C_W / k_2 [1 - \exp(-k_2 t)] \quad (1.3)$$

As time approaches infinity, the concentration in the fish at time t, $C_{F(t)}$, approaches the asymptote k_1C_W/k_2. Consequently, at equilibrium, the fish/water bioconcentration factor, BCF will be:

$$BCF = C_{F(t=\emptyset)} / C_W = (k_1 C_W / k_2) / C_W = k_1 / k_2 \quad (1.4)$$

Equation 1.2 has the following solution:

$$C_{F(t)} = C_{F(o)} \exp(-k_2 t) \quad (1.5)$$

often used in th-e log-transformed form:

$$\ln[C_{F(t)}] = \ln[C_{F(o)}] - k_2 t \quad (1.6)$$

When $C_{F(t)} = C_{F(o)}/2$, Equation 1.5 becomes:

$$C_{F(o)} / 2 = C_{F(o)} \exp(-k_2 t_{1/2}) \quad (1.7)$$

where $t_{1/2}$ is the period of time (h) needed to reduce C_F by 50%, or *half-life*. By means of a log-transformation, Equation 1.7 becomes:

$$\ln(C_{F(o)}) - \ln 2 = \ln(C_{F(o)}) - k_2 t_{1/2} \quad (1.8)$$

and

$$t_{1/2} = \ln 2 / k_2 \quad (1.9)$$

The time needed to obtain a 90% reduction of C_F, sometimes called $t_{0.1}$ (analogous to the expression $t_{0.5}$ for the half-life) or dissipation time 90, DT90 (Klein et al., 1993), is obtained by a similar procedure and is:

$$t_{0.1} = DT90 = \ln 10 / k_2 \tag{1.10}$$

In all these expressions, the units for t and rate constants k_1 and k_2 must be the same.

The example dealt with a simple monocompartmental first-order approach. More complex models are often needed to simulate real situations.

Once the model is described, formally defined, and applied, three further steps must be performed (Jørgensen, 1990):

- *verification:* the internal logic of the model and the physical or biological meaning of the selected parameters are tested.
- *calibration:* selected parameters are varied to obtain the best agreement with data produced by the simulation and field or laboratory data.
- *validation* of the quality of prediction produced by the model with respect to the simulated system is performed by examining the quality of fit between predicted and directly measured data.

Finally, it is important to know how the model reacts to changes in input data (essentially forcing functions or parameters). This is performed by the *sensitivity analysis.*

All these definitions are in agreement with the Standard Protocol for evaluating environmental-fate models of the American Society for Testing and Materials (ASTM), reported in Table 1.2.

In the following chapters, the focus will be on models simulating evaluative environments, rather than real environments *(evaluative models)*. The concept was introduced late in the 1970s by Baughman and Lassiter (1978), according to whom the environmental fate of chemicals may be investigated not only by means of very complex models, but also, and sometimes better, by these evaluative environments which are not a reproduction of any real environment, but of arbitrary systems. Thus, *evaluative models* refer to simplified systems.

> *Evaluative models have the advantage of being readily accessible to non-expert people; they are easy to run and easy to understand.*

The main limitation is that the results are not related to any real situation so that the absolute data may not be accurate. However, the relative output of one chemical with respect to another one may be applied to rank the potential of different substances to reach (and contaminate) selected environmental compartments.

Complex models for the environmental fate of chemicals have been available since the 1930s and a great number of new and improved models were produced,

Table 1.2. Definitions According ASTM (1984) for Evaluating Environmental-Fate Models

Term	Definition
Algorithm	The numerical technique embodied in the computer code
Calibration	A test of a model with known input and output information that is used to adjust or estimate factors for which data is not available
Compartmentalization	Division of the environment into discrete locations in time or space
Computer program	The assembly of numerical techniques, bookkeeping, and control language that represents the model from acceptance of input data and instructions to delivery of output
Model	An assembly of concepts in the form of a mathematical equation that portrays understanding of a natural phenomenon
Sensitivity	The degree to which the model result is affected by changes in a selected input parameter
Validation	Comparison of model results with numerical data independently derived from experiments or observations of the environment
Verification	Examination of the numerical technique in the computer program to ascertain that it truly represents the conceptual model and that there are no inherent numerical problems in obtaining a solution

particularly during the 1970s and 1980s. The problem is not in developing a more sophisticated and complete model, but when the model is applied: the deeper the physicomathematical description of the mechanisms determining the fate of the chemical in the environment, the greater is the difficulty in finding a value (and sometimes the meaning) for each parameter. It is essential to find an equilibrium between simplicity and complication. Too simple models could be too crude and misleading, while *an excessively detailed model is unlikely to be useful or even understandable* (Mackay, 1991).

1.3. THE ROLE OF PHYSICOCHEMICAL PROPERTIES

Physicochemical properties, intrinsic to the nature of different chemicals, are known to play a very important role in determining the environmental fate of contaminants. To a first approximation, for a given site and time and constant environmental variables, the driving forces of the environmental behavior of different substances essentially depend on their intrinsic properties.

In evaluative models, the environmental variables are often standardized, so that the influence of the physicochemical properties of the substances can be observed. For the purposes of contamination control, the question to be answered may be: "Among these chemicals, please indicate the worst" or "Rank the potential to reach and contaminate groundwater of these *n* substances." Now, these questions may refer to:

- a specified site
- a region

In the first case, standardization of the environmental properties is acceptable, at least in principle, since the site is the same. In the second case, site-specific variations in environmental properties can be overcome by regarding the whole region as having the environmental characteristics of the most vulnerable site.

Physicochemical properties can be divided into *molecular* and *molar properties*. Molecular properties are independent of state and are intrinsic to the single molecule. They include molecular structure, electronic charge, heat of formation, and heat of atomization. Molar properties depend on molecular aggregation; examples are density, molar mass, water solubility, melting and boiling points, and vapor pressure. Both play a significant role in regulating the environmental distribution, fate, and effects of contaminants.

1.3.1. Physical Properties

Molar mass (M): This property can be referred to any particle (atom, ion, formula, or molecule) and refers to the sum of the average atomic masses of all the atoms forming the particle, expressed in grams; so, the units of molar mass are grams per mole, as follows.

$$M = \mathrm{g/mol} \tag{1.11}$$

Molar mass is not synonymous with *molecular weight,* which is a molecular property that indicates the mass of a single molecule in *atomic mass units* (a.m.u.). The main application of molar mass is the normalization of mass units.

Heat of fusion (ΔHf): This property, also called latent heat of fusion or *enthalpy of fusion,* corresponds to the quantity of heat to be applied to a unit of mass of the substance at the fusion temperature to obtain the solid-to-liquid transition, with no temperature change (the energy is used to dissolve the crystals, typical of the solid state); units are cal/g or J/mol.

Heat of vaporization (ΔHv): Also called *enthalpy* (or latent heat) *of vaporization,* this property indicates the quantity of heat required to convert a unit mass of liquid into vapor, with no changes in temperature and pressure. This quantity decreases with increasing temperature, tending to zero as the critical temperature *(critical point)* is approached.

Heat of sublimation (ΔHs): This property also called *enthalpy* (or latent heat) *of sublimation,* indicates the heat required to convert a unit mass of solid into vapor, with no changes in temperature and pressure.

Melting point (mp): It is defined as temperature of the conversion from the solid state (crystals) to the liquid state. The *(normal) melting point, mp, indicates the temperature of the solid-to-liquid transition at a pressure of 101,325 pascal (Pa).* Melting point is related to the nature of the bonding forces of the solid. A low *mp* indicates relatively weak forces of attraction. Melting point has the dimensions of temperature and the units are degree Celsius (°C) or kelvin (K). The scale conversion is as follows:

$$K = {}^{\circ}C + 273.16 \quad (1.12)$$

The melting point indicates if, at a given environmental temperature, the substance is liquid or solid. This information is essential in comparing properties such as water solubility or vapor pressure of different substances.

Normal boiling point (bp): It is defined as *the temperature at which the vapor pressure of a pure compound in the liquid state reaches 101,325 Pa.* Boiling is a special form of evaporation, occurring when bubbles form in the liquid. The boiling point indicates whether the substance is, at a given temperature, in the liquid or gaseous phase. The boiling temperature can be modified by changing the pressure over the liquid or by adding other substances. For the purposes of the present discussion, the boiling point will indicate the normal boiling point. Beyond the boiling point, it is possible to identify a *superheated liquid* state for gases, which is analogous to the *subcooled liquid* state for solids (see Figure 1.4).

Trouton (1884) observed (see also Rechsteiner, 1990) that:

$$\Delta Hv / bp = 21 \text{ cal}/(\text{mol} \cdot \text{K}) = 88 \text{ J}/(\text{mol} \cdot \text{K}) \quad (1.13)$$

The relation in Equation 1.13 is known as *Trouton's rule.* Water and ethanol heat of vaporization values are a little higher than predicted by Trouton's rule, due to intermolecular forces which enhance the quantity of energy needed for vaporization. Chemicals with higher molecular weight and lower molecular interactions may have a slightly lower value of the constant. Mackay et al. (1986) applied a value of 84 J/(mol·K) for the ratio $\Delta Hv/bp$ to calculate the value of ΔHv from *bp* for organic chemical contaminants:

$$\Delta Hv = 84\, bp \quad (1.14)$$

where ΔHv is the heat of vaporization, in J/mol and *bp* the boiling point, K.

Critical point (cp): the maximum temperature in which liquid and gas phases of a chemical coexist in equilibrium; above critical temperature, the distinction between liquid and gas disappears and the substance exists as a *supercritical fluid* (Figure 1.4).

Vapor pressure (P): or equilibrium vapor pressure, is the pressure exerted by the vapor of a substance in equilibrium with its pure phase (liquid or solid) at a given temperature. Its dimensions are those of a pressure and the units are pascal (Pa). The conversion from other units can be made considering that one pascal is equal to one newton per square meter (Pa = N/m^2) and that the atmosphere (atm) and millimeters of mercury (mm Hg) are related to Pa as follows:

$$1 \text{ atm} = 101{,}325 \text{ Pa} \quad (1.15)$$

$$1 \text{ mm Hg} = 133.3 \text{ Pa} \quad (1.16)$$

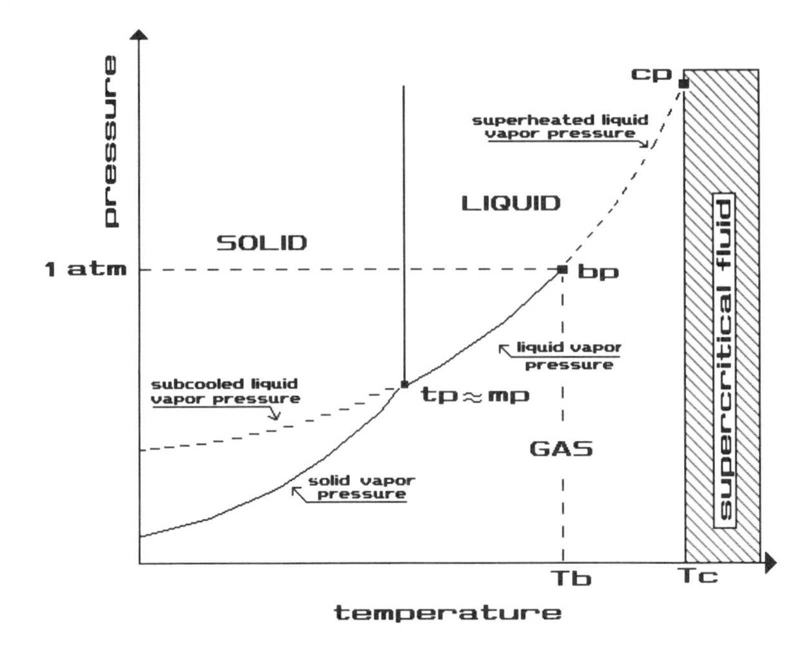

Figure 1.4. Simplified phase diagram for a pure substance; tp, triple point (indicating the equilibrium between the solid, liquid and gas phases); mp, melting point; bp, boiling point; cp, critical point; Tb, boiling temperature; Tc, critical temperature.

Vapor pressure is independent of the pressure of the system; in other words, the escaping tendency of a substance from its pure phase at a given temperature is the same at different elevations or even in absence of the atmosphere. Vapor pressure is a key property for determining the volatility potential of substances.

The dependance of vapor pressure on the physical state (solid or liquid) and on the temperature is shown in Figure 1.4. For a given temperature below the melting point, a solid substance has two different vapor pressure values: one as a solid, one as a subcooled liquid (Figure 1.4). Over short temperature ranges, Equation 1.14 can be applied to calculate the vapor pressure value at one temperature from the known value at another temperature by means of a form of the Clausius-Clapeyron equation (Mackay et al., 1986):

$$\ln\left(P_{S1}/P_{S2}\right) = (\Delta Hv/R)\left(1/T_2 - 1/T_1\right) = (84\ bp/R)\left(1/T_2 - 1/T_1\right) \quad (1.17)$$

where P_{S1} is the vapor pressure (Pa) of the solid at temperature T_1 (K), P_{S2} the vapor pressure of the solid at temperature T_2 (K), *bp* the boiling point (K), and R the gas constant, 8.314 Pa $m^3/(mol \cdot K)$. To check for units, it is important to remember that 1 J = 1 Pa m^3.

The simplest expression of the Clausius-Clapeyron equation, applicable for short variations in temperature when the enthalpy of sublimation or vaporization can be regarded as constant, is as follows:

Table 1.3. Parameters for the Clausius-Clapeyron Equation log (P_L) = A_L + B_L 1/T for Some Nonpolar and Semipolar Compounds. P_L is in Pa.

Compound	B_L	A_L
α-HCH	–3575	11.34
γ-HCH	–3680	11.15
p,p'-DDT	–4865	13.02
o,p'-DDT	–4626	12.77
p,p'-DDD	–4622	12.49
p,p'-DDE	–4554	12.79
Aldrin	–3924	12.04
Dieldrin	–4310	12.46
Mirex	–4718	12.27
Endosulfan I	–4201	11.87
Endosulfan II	–4306	12.08
Endosulfan sulfate	–4470	12.11
Endosulfan ether	–3704	11.23
Endosulfan lactone	–4077	11.65
Hexachlorobenzene	–3582	11.11
2,2′,5,5′-Tetrachlorobiphenyl	–4127	11.74
2,2′,4,5,5′-Pentachlorobiphenyl	–4369	12.13
2,2′,3,3′,5,5′,6,6′-Octachlorobiphenyl	–4851	12.99
Decachlorobiphenyl	–5402	13.27
Benzo[a]pyrene	–4989	11.59

From Hinckley et al. (1990). With permission.

$$\log(P_S) = A_S + B_S 1/T \tag{1.18}$$

for solids, and

$$\log(P_L) = A_L + B_L\ 1/T \tag{1.19}$$

for liquids. A and B are parameters (intercept and slope) depending on the chemical and on the temperature interval and T, the absolute temperature.

When P_L is in Pa, the parameters A_L and B_L can be obtained, within a factor of 2 of average literature values, by means of correlations with gas chromatographic retention data (Hinckley et al., 1990). Some examples are reported in Table 1.3.

For solids, the subcooled liquid vapor pressure, P_L, is generally higher than P_S (Figure 1.4). The value of P_L for the same compound can be calculated from P_S by means of the approach of the *fugacity ratio,* F, which is F = P_S/P_L (Suntio et al., 1988):

$$F = P_S / P_L = \exp\{-\Delta S_f[(mp/T) - 1]/R\} \tag{1.20}$$

where R is the gas constant, 8.314 Pa m^3/(mol·K), T the absolute temperature in K and ΔS_f the *entropy of fusion,* in J/(mol·K).

The entropy of fusion is the ratio between the heat of fusion and the melting point, ΔHf/mp.

In the absence of an experimental or calculated value of ΔS_f, to a first approximation, a standard vale of 56 J/(mol·K) can be adopted (Suntio et al., 1988). Average ΔS_f values of 56.5 J/(mol·K) have been reported by Yalkowsky (1979) for rigid aromatic molecules. The F values at 20°C can be obtained from Equation 1.21:

$$F = \exp[-0.023\,(mp - 293)] \tag{1.21}$$

The fugacity ratio values, F, can be applied to the calculation of the solubility in water of the substance as a liquid, S_L, from the solubility as a solid, S_S, being:

$$S_S / S_L = P_S / P_L = F \tag{1.22}$$

The solubility in water, S, *of a substance is the maximum amount of the chemical that dissolves in a given quantity of pure water at a selected temperature.* The dimensions generally are those of a mass/volume concentration, suitable units are: mol/m^3. This property is determinant in the environmental distribution and fate of contaminants: high solubility in water may favor escape from other environmental compartments, and migration to surface and groundwater.

In the environment, contaminants may occur in relatively large amounts, such as in the surface soil of a field treated with pesticides, or highly dispersed. In the first case, solids will conserve, for a certain time, crystalline aggregations (i.e., the solid state); in the second, a finely dispersed solid will behave as a *subcooled liquid.* The solubility in water of a pure solid compound is lower than its solubility as a subcooled liquid. For example, if we consider the two herbicides *butylate* and *atrazine*, they have water solubilities (at 20°C) of 0.18 and 0.14 mol/m^3, respectively. But butylate is liquid, while atrazine is solid, with a melting point of 174°C (Suntio et al., 1988). The solubility of butylate is higher than that of atrazine as it stands but, if finely dispersed, atrazine too will behave as a liquid, with a new solubility increased by a factor of 35.7, to 4.99 mol/m^3. The same happens when comparing the vapor pressure values: the vapor pressure of atrazine, as a solid, increases 35.7 times when considering the corresponding subcooled liquid (the fugacity ratio, F, is 0.028 and 1/0.028 = 35.7).

A rapid method of calculating the conversion factor from solid to liquid, similar to the method of the fugacity ratio, is that proposed by Yalkowsky et al. (1983), also applicable to vapor pressure (Suntio et al., 1988):

$$\log(S_L / S_C) = \log(P_L / P_C) = 0.01\,(mp - 25) \tag{1.23}$$

where S_L/S_C is the ratio between the solubility of the liquid and that of the corresponding solid; P_L/P_C the ratio between the vapor pressure of the liquid and that of the solid; *mp* the melting point in °C, and 25 is the working temperature (°C).

From Equation 1.23, it can be seen that a substance with a *mp* of 125°C has solubility and vapor pressure as a liquid 10 times that as a solid (and F = S_C/S_L

$= P_C/P_L = 0.1$). When *mp* is 225°C, the ratio $S_L/S_C = P_L/P_C$ is 100 (and F = 0.01). To a first approximation, each 100°C increase in melting point leads to an increase by a factor of 10 in the ratios S_L/S_C and P_L/P_C.

1.3.2. Chemical and Reaction Properties

Acid dissociation constant, K_a, and pK_a: A compound may be more or less ionized in water solution. Highly ionized organic and inorganic acids and bases behave in a very different way in the environment if compared with the corresponding neutral molecule. The degree of dissociation may greatly influence solubility in water, adsorption onto soil particles, and toxic potential. Dissociation is reversible; in the case of acetic acid, for instance:

$$CH_3COOH + H_2O \rightleftharpoons H_3O^+ + CH_3COO^- \tag{1.24}$$

In dilute solutions (e.g., <1 mol/m³, or <1 mmol/L), the *acid dissociation constant*, K_a, is:

$$K_a = \frac{[H_3O^+][CH_3COO^-]}{[CH_3COOH]} \tag{1.25}$$

The dimensions of K_a are those of a concentration and suitable units are mol/L. Since pH is the negative logarithm (base 10) of the hydrogen ion (or proton) concentration, in mol/L, $[H_3O^+]$:

$$pH = -\log[H_3O^+] \tag{1.26}$$

from Equation 1.25, taking the negative logarithm of both sides:

$$-\log K_a = -\log\frac{[H_3O^+][CH_3COO^-]}{[CH_3COOH]} \tag{1.27}$$

and, for dilute solutions:

$$-\log K_a = pH - \log\frac{[CH_3COO^-]}{[CH_3COOH]} \tag{1.28}$$

Calling the negative logarithm of K_a, as for pH, pK_a:

$$pK_a = pH - \log\frac{[CH_3COO^-]}{[CH_3COOH]} \tag{1.29}$$

or

$$\log\frac{[CH_3COO^-]}{[CH_3COOH]} = pH - pK_a \tag{1.30}$$

The pK_a is a measure of the *strength* of an organic acid relative to the acid/base pair $[H_3O^+]/[H_2O]$; pK_a indicates at which hydrogen ion activity the substance is 50% dissociated: when pH is equal to pKa, the concentration of the acetic acid in the dissociated form $[CH_3COO^-]$ and in the undissociated $[CH_3COOH]$ are equal. This happens, for acetic acid, at pH 4.75, being its pK_a = 4.75. From Equation 1.30, at pH 3.75 the acetic acid will be 9.1% dissociated, at pH 2.75 about 1% dissociated, while at pH 5.75 it will be dissociated at 90.9%, 99% at pH 6.75, 99.9% at pH 7.75, etc. When an organic acid is dissociated, for instance in a water body, this implies that it is in the form of an anion (A^-).

In the case of basic compounds, taking into account their conjugated acids, a similar relationship as that in Equation 1.30 is easily developed (Harris and Hayes, 1990), but with an inverse trend: when pH = pK_a, the compound is still 50% dissociated, but increasing the pH value the compound becomes less and less dissociated. For instance, *atrazine*, an *s*-triazine herbicide, is a weak base with pK_a = 1.68 (Weber, 1970) for the conjugated acid (protonated form; Welhouse and Bleam, 1993). So, at pH = 7.68, atrazine is 99.9999% undissociated, or in the neutral form (Figure 1.5).

For bases, a basic dissociation constant, K_b, for the reaction with water can be defined as follows:

$$B + H_2O \rightleftharpoons BH^+ + OH^- \tag{1.31}$$

with

$$K_b = \frac{[BH^+][OH^-]}{[B]} \tag{1.32}$$

The reaction of the base with water produces a cation, BH^+, which is the corresponding conjugated acid. The reaction of the conjugated acid with water is as follows: $BH^+ + H_2O \rightleftharpoons H_3O^+ + B$ and the acid dissociation constant for the chemical species BH^+ is:

$$K_a = \frac{[B][H_3O^+]}{[BH^+]} \text{ and } pK_a = pH - \log\{[B]/[BH^+]\} \tag{1.33}$$

The relationships between degree of ionization and pH can be resumed as follows (Henderson-Hasselbach equations):

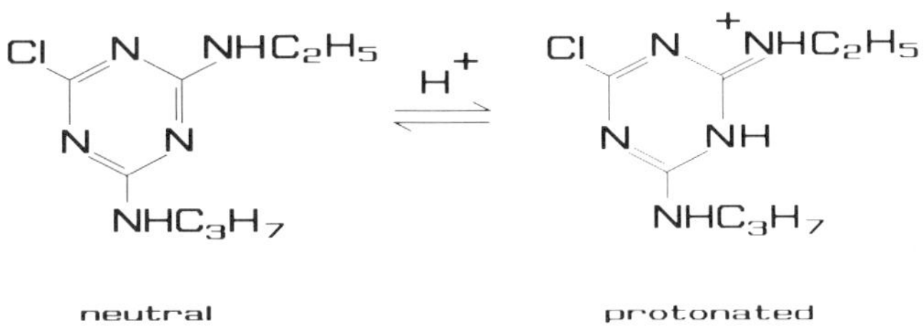

neutral protonated

Figure 1.5. Neutral and protonated forms of atrazine.

$$\text{for acids: } \log(\text{nonionized/ionized}) = pK_a - pH \quad (1.34)$$

$$\text{for bases: } \log(\text{ionized/nonionized}) = pK_a - pH \quad (1.35)$$

The polarity of the chemical will greatly influence its environmental distribution, as for instance, the soil/water partition and route of absorption by living organisms. Consequently, the mobility of a chemical from soil to water will greatly depend on its electrical charge: in general, soils are negatively charged, so positive ions (e.g., the herbicide *paraquat*) are easily retained by soils, while negative ions (Cl^-, NO_3^-, SO_4^{2-}) are able to reach groundwater; nonpolar chemicals will be retained in a proportion inverse to their solubility in water.

A chemical may be more or less *conservative* under environmental conditions. Conservative chemicals are refractory to reactions, maintaining their structure and properties for a long time. If sufficiently mobile, they are typical *global contaminants*. Less conservative chemicals may become global contaminants only if their input (global input) is able to compensate their destruction.

Environmental chemical reactions include hydrolysis, photolysis, oxidation, reduction, and biodegradation.

Hydrolysis is a reaction with water and, consequently, one of the most important processes. For organic chemicals, RX, hydrolysis consists in a displacement of the X by a carbon-oxygen bond, as follows:

$$R-X \xrightarrow{H_2O} R-OH + X^- + H_3O^+ \quad (1.36)$$

This transformation of the organic substance leads to an increase in polarity. The kinetics of hydrolytic reactions in water may be expressed as *first-order* in the concentration of RX (Mill, 1980):

$$d[RX]/dt = -k_{hy}[RX] \quad (1.37)$$

The first term of the equation indicates the variation of the concentration of the chemical species RX (mol/m^3) with time, t (h); this quantity is the *rate of hydrolysis*. The sign "–" in the second term indicates that this variation consists of a reduction of the concentration of RX. The rate of this reduction, or *hydrolysis rate*, is directly proportional to the concentration of RX (mol/m^3) and to a *rate constant* k_{hy}, always positive, which is the *hydrolysis rate constant* (dimensionally a frequency: 1/t and here 1/h).

Equation 1.37 refers to the change, with time, of the *concentration* of the chemical RX. If the change in *quantity* of the chemical RX, Q_{RX} (mol), has to be represented in the same process, the following relationship may be used:

$$dQ_{RX} / dt = -V\, k_{hy}[RX] \quad (1.38)$$

where V is the volume (m^3) of the solution containing RX at the concentration [RX], mol/m^3.

In Equation 1.38, the rate constant k_{hy} indicates the fraction of chemical reacting in the unit of time: for instance, a constant of 0.1 h^{-1} means that the mass fraction of chemical reacting in 1 h is 0.1; in other words, this means that the reaction is going on transforming 10% of the chemical per hour. One of the properties of this type of kinetics is that: *while the rate of the reaction is progressively slowing down, the fraction of chemical reacting is constant and not depending on the chemical concentration or quantity.* This leads to a constant time to reduce the concentration or the quantity to half of the initial one; this time (or better *period*) is currently called *half-life,* symbol $t_{1/2}$, (example of units: hours). The half-life, $t_{1/2}$ is inversely proportional to the rate constant k_{hy} and can be calculated as follows (see also Equation 1.9):

$$t_{1/2} = \ln 2 / k_{hy} \quad (1.39)$$

where ln is the logarithm base *e* (*e* = 2.7183; ln 2 = 0.693). If the chemical is reacting with a rate constant of 0.1 h^{-1} (which is 10% per hour), the time required to halve a given concentration will be 0.693/0.1 = 6.93 h. In studies carried out to measure the rate of hydrolysis, it is important to follow the experiment for at least two half-lives, in order to reduce the initial quantity to about 1/4 of itself.

Equations 1.37 and 1.38 are typical *working equations* with precise physical meanings. However, for practical purposes, some mathematical manipulations are needed so that they can be more effectively applied to experimental data: from Equation 1.37, by separation of the variables RX and t:

$$d[RX]/[RX] = -k_{hy}\, dt \quad (1.40)$$

Integrating:

$$\ln[RX] = -k_{hy} t + \text{constant} \quad (1.41)$$

where "constant" means the *integration constant,* which is in this case, the logarithm of the value of [RX] at the initial time (t_o), or $[RX]_o$; consequently,

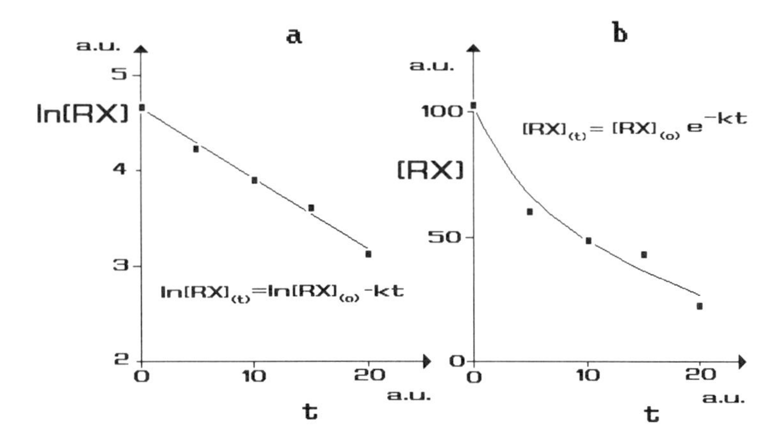

Figure 1.6. Semilog (a) and normal (b) plots of a first order clearance process. a.u. = arbitrary units.

$$\ln[RX] = \ln[RX]_o - k_{hy}t \tag{1.42}$$

Equation 1.42 is a straight line (Y = a + bX), with Y axis *intercept* (a) equal to $\ln[RX]_o$ and *slope* $-k_{hy}$ (t is the independent variable). So by plotting ln[RX] vs t, referring to experimental data, a validation of the correctness of the assumption of the first-order kinetic model is possible. An example is given in Figure 1.6a. The calculation of the parameters of the function in Figure 1.6a can be performed by current statistical instruments, such as the *least-squares* method. Figure 1.6b represents the same decay function (in analogy with the *decay* of radioactive elements) without a semilog plot, according to the following relationship derived from Equation 1.42, by eliminating logarithms from both sides:

$$[RX]_{(t)} = [RX]_{(o)} \exp(-k_{hy}t) \tag{1.43}$$

If the decay function needs to be applied to the quantity Q_{RX}, as shown in Equation 1.38, rather than to the concentration of the substance, [RX], with a similar approach the following equation can be obtained:

$$Q_{RX(t)} = Q_{RX(o)} \exp(-k_{hy}t) \tag{1.44}$$

The transformation from Equation 1.42 to Equation 1.43 may introduce a bias due to the omission of the *random error term* in the first Equation and the consequent omission of its transform in the second (Newman, 1993). More details on this particular, but important, topic are discussed in Appendix 1. The key element in these reaction kinetics is the rate constant k_{hy}, which contains all the information on the stability of the substance under given conditions. A low value of k_{hy}, for example 10^{-5} h^{-1}, will indicate a high stability to hydrolysis (0.00001 is the mass fraction of chemical reacting in an hour).

Photolysis in water, on soil surfaces, or in the atmosphere is another reaction mechanism which can be significant in determining the environmental fate of contaminants. The process depends on the intensity of light *(irradiation intensity)* and on the wavelength. The solar spectrum cut-off by the upper atmosphere is at about 290 nm. The absorption of light of this or a longer wavelength by a molecule generates an excitation, which may produce decay back to its ground state (no reaction) or chemical transformation (reaction). Under environmental sunlight conditions, photochemical transformation may occur, particularly by the action of light in the wavelength range 290–400 nm, which contains energies comparable to chemical bond energies. The energy of a quantum of light (or photon), E, in kilojoules (kJ), is related to the frequency of light by the following relationship:

$$E = h\nu \tag{1.45}$$

where h is the Planck's constant (6.626×10^{-34} J s) and ν is the frequency of light, in s^{-1}. The frequency of light is given by:

$$\nu = c / \lambda \tag{1.46}$$

where c is the speed of light (3×10^{10} cm/s) and λ is the wavelength, in cm. Consequently, combining the last two equations:

$$E = hc/\lambda \tag{1.47}$$

The energy per mole of photons, in J/einstein (one mole of photons is an *einstein*), can be calculated from the relationship:

$$E = Nhc / \lambda \tag{1.48}$$

where N is the Avogadro's number (6.022×10^{23} molecules/mol), h is the Planck's constant (6.626×10^{-34} J s), and c is the speed of light (3×10^{10} cm/s).

From Equation 1.48, light with a wavelength of 300 nm (or 3×10^{-5} cm) contains an energy, E, of 399,018 J/einstein (or J/mol of photons). E is currently expressed in kJ/einstein, sometimes in kcal/einstein, and λ in nm. A practical formula to calculate E from λ is as follows (Leifer, 1988):

$$E \text{ (kJ / einstein)} = 1.20 \times 10^5 / \lambda \text{ (nm)} \tag{1.49}$$

$$E \text{ (kcal / einstein)} = 2.86 \times 10^4 / \lambda \text{ (nm)} \tag{1.50}$$

The energy of the light is inversely proportional to the wavelength (i.e., doubling the wavelength, the corresponding energy is one half): at λ = 600 nm, the corresponding E is about 200 kJ/einstein.

Table 1.4. Bond Energies in Various Organic Compounds

Bond	Compound group	Bond energy (kJ/mol)
C-H	Alkane, alkene, alkyne, arene	322-523
C-C	Alkane	260–456
C-O	Ester, alcohol, ether	322–406
C-N	Amine, amide	276–414
C-S	Thio	272
C-F	Alkyl	439–452
C-Cl	Alkyl, aryl	251–435
C-Br	Alkyl, aryl	197–143
C-I	Alkyl	205–234
O-H	Alcohol, phenol	456–465
N-H	Amine, amide, imine	385–402
O-O	Peroxide	142–184
N-N	Azine	155–163
S-S	Disulfide	226

From Cessna and Muir, 1991. With permission.

Energies of light in the spectral range of interest for environmental photochemistry (290-800 nm) are of the same order of magnitude as bond energies found in various organic compounds (Table 1.4). Chemicals which are refractory to other reaction mechanisms may easily undergo photoreaction. As an example, methylmercury is quite a stable compound in the dark (as under aquatic sediments or inside the body of opaque organisms); however, in environmental compartments reached by sunlight, it may be easily transformed into inorganic mercury by photolysis: the mean dissociation energy of the C-Hg bond is only 120 kJ/mol (Mortimer, 1962).

Reactions have been grouped in two categories: *direct* photolysis and *sensitized photolysis* (Zepp et al., 1978). The former is initiated via direct light absorption by the chemical under discussion; the latter needs an intermediate compound, which absorbs light directly and is called a *photosensitizer*. Direct photoreaction potential indicates the background photoreactivity of the chemical, while the indirect processes are more site specific.

To the photoreaction of microcontaminants (both by direct and indirect photolysis) in the environment, first-order kinetics may be applied (Leifer, 1988):

$$d[\mathrm{C}]/dt = -\mathrm{k}_{\mathrm{ph}}[\mathrm{C}] \tag{1.51}$$

where the overall photoreaction rate constant k_{ph} (t^{-1}) is:

$$\mathrm{k}_{\mathrm{ph}} = \mathrm{k}_{\mathrm{d}} + \mathrm{k}_{\mathrm{s}} \tag{1.52}$$

where k_d and k_s indicate, respectively, the direct and indirect photoreaction rate constants.

Oxidation and *reduction (redox)* processes are another important reaction mechanism in the environment. Molecular oxygen is a weak oxidant: most organic compounds do not react spontaneously at significant rates with this chemical species. Oxidations of organic contaminants are usually mediated by free radicals, ozone, and singlet oxygen. Oxygen radicals (O·, or singlet oxygen), are formed by photosensitized reactions and may react with double bonds to form peroxides. Free radicals are highly reactive chemical species characterized by a very short lifetime (on the order of a billionth of a second) and may activate chain reactions. In aquatic environments peroxy ($RO_2\cdot$) and alkoxy (RO·) radicals may give rise to oxidation processes as follows:

$$RO_2\cdot + RH \rightarrow RO_2H + R\cdot \tag{1.53}$$

and

$$RO\cdot + RH \rightarrow ROH + R\cdot \tag{1.54}$$

In the atmosphere, hydroxyl radicals (OH·) may react with clorofluorocarbons, producing chlorine compounds and chlorine radicals (Cl·); chlorine radicals are also produced by the photoreaction of clorofluorocarbons and molecular chlorine in the higher troposphere and are involved in stratospheric ozone destruction processes.

Reduction reactions may also be involved in the chemical transformation of organic contaminants. In fact, most synthetic organic compounds are designed to survive in aerobic environments (Schwarzenbach et al., 1993). However, under anaerobic conditions, some classes of compounds may undergo reduction. Anaerobic sediments contain natural reductants, such as reduced inorganic forms of iron and sulfur (e.g., iron (II) sulfides, hydrogen sulfide), which are able to react with reducible organic contaminants (such as the nitrobenzenes), by means of electron transfer mediators (Schwarzenbach et al., 1990).

First-order kinetics can also be applied to oxidation and reduction processes associated with chemical trasformation of trace contaminants (i.e., pesticide residues), leading to an overall oxidation (k_{ox}) or reduction (k_{re}) rate constant.

Biodegradation refers to the breakdown of organic chemicals due to biological attack. Biodegradation may be *complete* when the organic chemical is converted to inorganic compounds, or *partial* when a structural change in the parent compound is produced. This is also called *primary biodegradation* and is of great interest in environmental chemistry. All living organisms may promote biodegradation processes; however, the major role in the environment is played by microorganisms. Biodegradation occurs by means of different mechanisms. These include oxidation and reduction, conjugating, and hydrolysis. One of the differences from other non-biological reactions concerns the reaction kinetics, generally faster than in abiotic systems. The model used to describe reaction rates in biodegradation is derived from the hyperbolic rate law introduced by Monod in the 1940s (see Scow, 1990), and currently applied to express the growth of microbial populations:

$$V = \frac{V_{max}\,[Su]}{K_m + [Su]} \tag{1.55}$$

where V and V_{max} indicate, respectively, the specific and maximum growth rate of microorganisms; [Su] the concentration of the substrate; and K_m the value of [Su] supporting half-maximum growth rate ($V_{max}/2$). With a low concentration of substrate, as is the case when the substrate is a microcontaminant, an expression for the biodegradation rate of the substrate can be derived from Equation 1.55 (Scow, 1990):

$$d[Su]/dt = -k_{bi}\,[Su][N] \tag{1.56}$$

where [N] is the concentration of microorganisms. From Equation 1.56, biodegradation appears to follow second-order kinetics, the rate depending on both the concentration of the chemical and the concentration of the microorganisms. Under current environmental conditions, however, the number of microorganisms effecting biodegradation is often independent of the concentration of the substrate (i.e., the chemical undergoing degradation). In this case, Equation 1.56 can be written as a first-order reaction:

$$d[Su]/dt = -k_{bi}[Su] \tag{1.57}$$

When all reactions are assumed to follow first-order kinetics, in a complex system where different degradation processes are occurring at the same time, it is possible to obtain the *first-order overall reaction rate constant,* k_R, by simply adding all rate constants:

$$k_R = k_{hy} + k_{ph} + k_{ox} + k_{re} + k_{bi} \tag{1.58}$$

Consequently, the *overall half-life,* $t_{1/2\,ov}$, is:

$$t_{1/2\,ov} - \ln 2 / k_R \tag{1.59}$$

and a residence time, T, can be defined as:

$$T = 1/k_R \tag{1.60}$$

In a laboratory model, it is possible to measure k_R under controlled conditions. When using open systems, it is important to avoid losses by mass transport (via water or air) influencing results. Figure 1.7 illustrates the decay of cypermethrin, a synthetic pyrethroid insecticide, in spiked soils at two different levels of con-

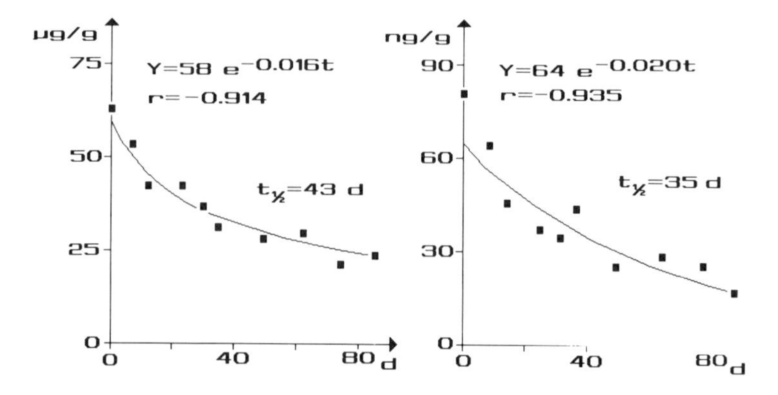

Figure 1.7. Decay of cypermethrin in spiked soils at two different levels of contamination. (From Bacci et al., 1987. With permission.)

tamination. Experimental data refer to the disappearance of the chemical but, the mass transport being absolutely negligible, in this case disappearance is only due to chemical reaction.

This kind of information is very useful in ecotoxicology: a simple laboratory model can supply homogeneous data for the overall degradation rate in soil for different chemicals. The findings from these models may differ from the field, where variations with space and time may occur. However, within a list of substances, these tests can lead to a ranking of chemical stability in different environmental matrices: absolute data may not be completely accurate, but the ranking has enough probability to be sufficiently precise.

Literature data and calculation by estimation methods (Lyman et al., 1990), together with direct measurements, represent the means to obtain information on chemical reaction rates.

1.3.3. Persistence: Reactivity and Mobility

The adjective *persistent* is often used to mean *refractory to chemical transformations.* This can be misleading: in some conditions, even the very ephemeral substances can be very persistent. To understand environmental chemistry, good definitions are essential. *Persistence* is a physical entity, with dimensions of time and consequent units (hours, days, years, etc.); it is synonymous with *residence time* and *turnover time*. Persistence can only be defined under *steady-state* conditions, that is, when the total quantity of the substance under study in a selected environmental compartment does not vary with time *(all time derivatives are zero)*. This implies that input and output are equal (Figure 1.8).

If I and O, the input and output rates, are mol/h, Q is mol, T is the persistence (in hours) and indicates the time needed for entering into the system a quantity of the chemical equal to the quantity contained in the system. As indicated in Figure 1.8, it is important to observe that the reciprocal of T is a frequency, 1/t, indicating the input or output rate constants which are analogous to the first-order reaction

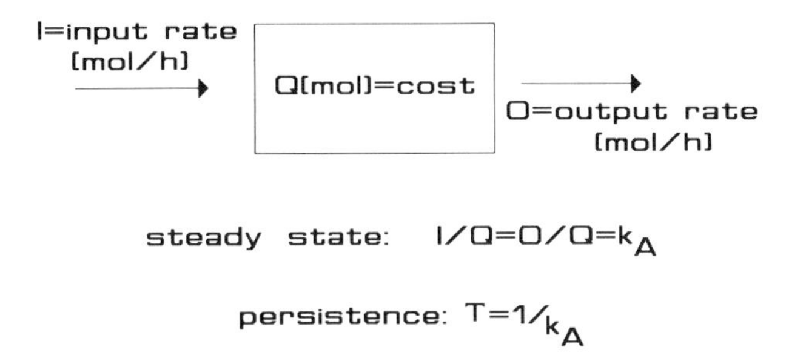

Figure 1.8. Steady state and persistence.

rate constants, discussed previously: the disappearance of a chemical from a compartment may happen not solely by chemical reaction processes, but also by mass transport *(advection)*,* essentially via air or water turnover.

If the simple model in Figure 1.8 is applied to a lake containing 10 mol of the conservative substance A, with I = O = 0.5 mol/h, the persistence of A is 10/0.5 = 20 h, and the value of the input- and output-rate constants, k_A, is 0.5/10 = 0.05 h^{-1}. In this case it is assumed that no reaction occurs, so the output rate constant can be defined as *advection rate constant, k_A, indicating the mass fraction of chemical A removed from the lake in the unit of time* (in the example 0.05, corresponding to 5% per hour). In analogy with reaction rate constants, it is also possible to calculate an advection half-life, as ln $2/k_A$, indicating the time needed to reduce the quantity of a chemical in the compartment by 50%, after stopping the input.

It is important to observe that similar calculations can be made with Q as the volume of the water in the lake (m^3), I and O in m^3/h. In this case, k_A indicates the volume fraction of the lake water removed in 1 h and T the persistence or the turnover time of the water in the lake.

Finally, it is possible to combine, under steady-state conditions, not only all reaction processes by means of the overall reaction rate constant, k_R, but also *mobility*, by means of the advection constant, k_A, defined above. This will lead to a first-order *disappearance rate constant,* k_{dis}:

$$k_{dis} = k_R + k_A \tag{1.61}$$

The overall persistence (or residence time, or turnover time) of a chemical in an environmental compartment assumed to be in a steady state is:

$$T = 1/(k_R + k_A) = 1/k_{dis} \tag{1.62}$$

The disappearance half-life will be $\ln 2/k_{dis}$.

* Advection can be defined as the transport of a chemical from an environmental compartment by bulk flow.

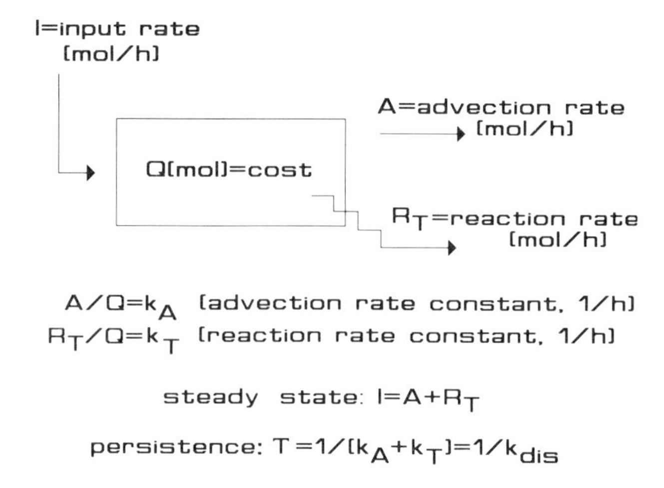

Figure 1.9. Steady state with advection and reaction.

Figure 1.9 shows an example representing an environmental compartment (e.g., the water of a lake) where there is a constant mass, Q, of a chemical maintained by the equilibration input-output, with advection and reaction. The persistence of a substance in a given environmental compartment depends on its chemical stability and, jointly, on the renewal by mass transport. To apply persistence properties to a system, it is essential to assume steady-state conditions. *The persistence concept is also related to the time for dissipation in water or air when the steady state is referred to the fluid.* Let us take the air of a small greenhouse, with a volume $V = 0.2\ m^3$ and air flowing in and out at an unknown rate: it is possible to evaluate the air turnover time, or persistence in the air inside the greenhouse (the system is in steady-state for the air mass: I = O) by means of the *dissipation kinetics* of a tracer. If we load the air with γ-hexachlorocyclohexane (γ-HCH) vapors, by simply opening a little bottle containing 1 g of pure chemical inside the greenhouse, previously sealed, then we open the system to allow the air to flow in and out, the concentration, C, of γ-HCH will decrease following first-order kinetics:

$$dC/dt = -kC \tag{1.63}$$

or

$$C_{(t)} = C_o \exp(-kt) \tag{1.64}$$

where *k is the dissipation rate constant* (1/h) and is numerically and dimensionally equal to the air advection constant, assuming no chemical reaction and no adsorption on the walls of the empty greenhouse. In a system like this, Bacci and Gaggi (1986), measuring C at different time intervals, found a value of $k = 0.75\ h^{-1}$, indicating that the dissipation was occurring at 75% of the quantity of γ-HCH present in the greenhouse air per hour. This also means that the air was flowing out (and in) with the same rate constant $k_A = 0.75\ h^{-1}$ and that the persistence of the air in the system, T, was:

$$T = 1/k_A = 1.33h\ (80\ min) \quad (1.65)$$

Once the persistence was known, from the greenhouse volume ($0.2\ m^3 = 200\ L$) the air in- and outflow rate was obtained: V/T = 200/80 = 2.5 L/min. A similar approach can be applied to a water reservoir where a contaminant is well mixed with water (as in a continuously stirred tank reactor).

1.3.4. Partition Coefficients

If *steady state* means no change in the system properties with time, it could be said that steady state and *equilibrium* are, in environmental chemistry, synonymous: a system in equilibrium also means no change in concentrations or quantities with time. The difference is that, for instance, *in a two-phase system, the equilibrium is reached when the ratio of concentrations of the chemical is equal to the partition coefficient,* while steady state does not imply reaching an equilibrium condition.

A partition coefficient of a chemical between phase 1 and 2, K_{12}, is:

$$K_{12} = C_1 / C_2 \quad (1.66)$$

where C_1 and C_2 indicate the concentration of the substance in the two phases in equilibrium. As a ratio of concentration, if these are expressed in homogeneous units (e.g., mol/m^3, g/L, mg/kg), K_{12} is typically a pure number, which means *dimensionless.* It is very important to remark that, even in a dimensionless form, *the partition coefficient keeps the memory of its origin* and may have different expressions and, consequently, different meanings. An illustration of this concept is given in Figure 1.10.

1.3.4.1. K_{OW} and K_{AW}

The most "famous" partition coefficient is the *1-octanol/water partition coefficient,* K_{OW} (in some previous works also called P or P_{OW}): this can be defined as the *ratio of a chemical's concentration in 1-octanol to its concentration in water, at equilibrium, in a closed system composed of octanol and water.* It is dimensionless, but it originates from a ratio of two concentrations, both expressed in mass/volume units (e.g., mol/m^3). If the volumes of octanol and water are equal, K_{OW} = 10 means that if we have 11 parts (in mass), 10 are distributed in the octanol and 1 in the water phase; K_{OW} = 1,000,000 means that over 1,000,001 parts, in mass units, 1,000,000 are in the octanol and 1 in the water. When the volumes of octanol

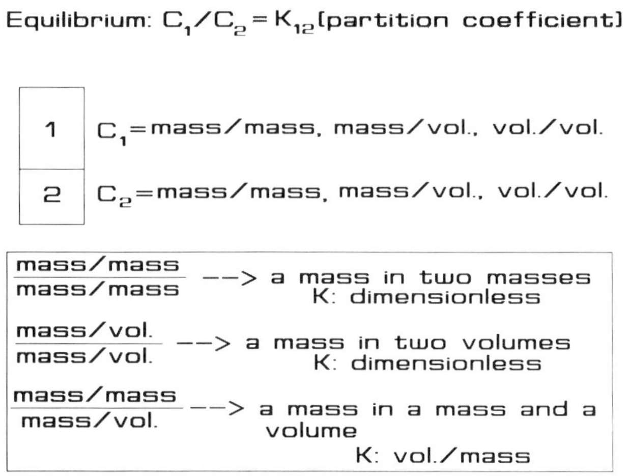

Equilibrium: $C_1/C_2 = K_{12}$ [partition coefficient]

1	C_1 = mass/mass, mass/vol., vol./vol.
2	C_2 = mass/mass, mass/vol., vol./vol.

$\frac{\text{mass/mass}}{\text{mass/mass}}$ --> a mass in two masses
K: dimensionless

$\frac{\text{mass/vol.}}{\text{mass/vol.}}$ --> a mass in two volumes
K: dimensionless

$\frac{\text{mass/mass}}{\text{mass/vol.}}$ --> a mass in a mass and a volume
K: vol./mass

Figure 1.10. Examples of different expressions and meanings for partition coefficients.

and water are different, the mass will be distributed to reach a concentration ratio equal to K_{OW}. Currently, the decimal logarithm (base 10 logarithm) of K_{OW} is used: log K_{OW}.

For organic chemicals, K_{OW} ranges from 10^{-3} and 10^{7} (and log K_{OW} from –3 and 7). Octanol simulates biological lipids and K_{OW} is applied to evaluate the lipid/water partition coefficient. Due to the relatively high reciprocal solubility of water and octanol (4.5 mol/m^3 of octanol in water phase and 2300 mol/m^3 of water in octanol; Leo and Hansch, 1971), K_{OW} does not simply indicate the ratio of the solubility in octanol and in water of a selected compound.

The solubility in octanol of organic substances is less scattered than respective water solubility, and tends to be rather constant in the range 200 to 2000 mol/m^3. Several organic chemical have the same lipophilicity, but they may differ in water solubility; consequently, K_{OW} is a measure of hydrophobicity, rather than a measure of lipophilicity (Mackay, 1991). *When log $K_{OW} > 3$, substances are considered hydrophobic.*

The K_{OW} values can be found in manuals and recent publications. For pesticides, for instance, the last version of the famous Pesticide Manual (Worthing and Hance, 1991) indicates K_{OW} values for the majority of the active ingredients. An alternative to existing literature data is direct measurement. Estimation methods are also available. As an example, the correlation developed by Yalkowsky et al. (1983), based on the dependence of K_{OW} on water solubility, may be applied to nonpolar aromatic compounds:

$$0.944 \log K_{OW} = -\log S_C - 0.01\, mp + 0.323 \tag{1.67}$$

where S_C is the water solubility in the solid state (in mol/L) and *mp* the melting point (°C). For liquids, the relationship is applied based on a value of *mp* = 25°C.

Table 1.5. Relationship Between Melting Point (*mp*), Water Solubility (S), and K_{OW}

Solute	*mp* (°C)	log K_{OW}	log S (as solid) (mol/L)
Anthracene	216	4.55	–6.56
Phenanthrene	101	4.55	–5.21

From Yalkowsky et al., 1983. With permission.

There is an important difference between K_{OW} and S; the latter is influenced by crystal interaction energy effects, leading to two different S values—one for the solid, the other for the corresponding subcooled liquid. On the other hand, K_{OW}, as other partition coefficients, is the same for the solid or liquid states: in both phases, crystal interactions contrast the solubilization to the same extent, leading to the same concentration ratio. An interesting example is reported in Table 1.5.

By means of Equation 1.23, the value of S_L (as a subcooled liquid) can be calculated for anthracene and phenanthrene: it can be seen that, while the increase from S_C to S_L is by a factor of 81.28 for anthracene, this factor is reduced to only 5.75 for phenanthrene; values of S_L are 2.25×10^{-5} and 3.55×10^{-5} mol/L, respectively. These values, in the same order of magnitude, are in agreement with equal log K_{OW} (Table 1.5).

Another very useful partition indicator is the air/water partition coefficient, K_{AW}, "offspring" of the more applied, in the past, Henry's law constant, H. The main problem in applying H is that it has dimensions of pressure per vol/mass (e.g., Pa m^3/mol) and its numerical meaning is not immediately comprehensible, particularly when comparing H values espressed in different units. On the other hand, K_{AW} is the *ratio, at equilibrium, of a chemical concentration in air to its concentration in water*, both expressed in mass/vol units (e.g., mol/m^3). Consequently, K_{AW} is dimensionless and analogous to K_{OW}.

While K_{OW} may be obtained by direct measurements, K_{AW} is currently calculated by the ratio "solubility in air"/solubility in water, S. The solubility in air of a substance, mol/m^3, can be calculated from its vapor pressure as follows:

$$\text{Solubility in air}\left(\text{mol/m}^3\right) = P/RT \tag{1.68}$$

where P is the vapor pressure, in Pa; R is the gas constant or 8.314 Pa m^3/(mol·K); and T the absolute temperature (K). Consequently,

$$K_{AW} = \frac{P/(RT)}{S} \tag{1.69}$$

It is important that both P and S refer to the same physical aggregation state (solid or liquid) and to the same temperature in order to avoid improper applications.

To convert H into K_{AW}, the following relationship may be applied (25°C):

$$H\left(\text{atm}\cdot\text{m}^3/\text{mol}\right)\times 41 = K_{AW} \quad (1.70)$$

The meaning of K_{AW} is analogous to that of K_{OW}: a K_{AW} of 0.3 indicates that if we have a concentration in the air of 3 mol/m^3, in the water, at equilibrium, we have 10 mol/m^3. With equal volumes of air and water, the same K_{AW} value indicates that with 13 parts (mass), 3 are distributed in air and 10 in water. The K_{AW} of water is of particular significance: at 20°C, it is 1.7×10^{-5}, indicating that the equilibrium concentration of water in air and water are directly proportional, respectively, to 1.7 and 100,000. Chemicals with K_{AW} two orders of magnitude, or more, below that of water will be trapped in this phase, with no possibility of reaching air before the water is eliminated. Chemicals with K_{AW} of 10^{-5} and more are able to pass from water to air.

1.3.4.2. K_{OW}-Related Partition Coefficients

Other important partition coefficients, quantitatively related to K_{OW}, are the soil (or sediment) organic carbon/water, K_{OC}; the soil or sediment/water, K_{PW}; the fish/water, K_{FW}; and the leaf/air, K_{LA}. The work by Karickhoff (1981) demonstrated that, for hydrophobic organic chemicals, the sorbing capacity of soil and sediments was controlled by the organic carbon fraction and that the organic carbon partition coefficient, K_{OC}, was related to K_{OW} as follows (Mackay, 1991):

$$K_{OC} = 0.41\, K_{OW} \quad (1.71)$$

As this is an empirical correlation, derived from experimental measurements of K_{OC} where the concentration in soil or sediment was expressed in milligrams per kilogram (mg/kg) and that in water in milligrams per liter (mg/L), K_{OC} has dimensions of a reciprocal density (i.e., volume/mass) and consequent units (e.g., L/kg). To obtain a dimensionless K_{OC} similar to K_{OW} it is necessary to multiply K_{OC} in Equation 1.71 by the sediment or soil bulk density, ρ_S, in kg/L or equivalent units. With a soil bulk density of 1.5 kg/L,

$$\text{dimensionless } K_{OC} = \rho_S\, K_{OC} = 1.5 \times 0.41\, K_{OW} = 0.615\, K_{OW} \quad (1.72)$$

indicating that the organic carbon in soils and sediments behaves as it was about 60% 1-octanol.

Calling K_p the soil or sediment/water partition coefficient, for organic nonpolar chemicals:

$$K_p = K_{OC}\, f_{OC} \quad (1.73)$$

where f_{OC} is the mass fraction of organic carbon in the soil or sediment. For instance, soil containing an organic carbon mass fraction = 0.01 (1% of organic carbon) will show a $K_p = 0.01\ K_{OC}$. The soil/water partition coefficient K_p will have the dimensions of selected K_{OC}: a reciprocal of a density (vol/mass) or dimensionless. When K_p has the dimensions of a reciprocal density, to obtain a dimensionless sediment/water partition coefficient, K_{PW}, analogous to K_{OW} and K_{AW}, K_p needs to be multiplied by the sediment, or soil, or particulate density, ρ_s (kg/L):

$$K_{PW} = \rho_s\ K_p \tag{1.74}$$

In the case of aquatic organisms, the "partition" of substance with water may be generated by three different mechanisms (Neely, 1980):

- *Bioconcentration:* is the intake exclusively via respiration, from water in aquatic systems, from air in terrestrial environments.
- *Bioaccumulation:* or the intake of a chemical by a living organism by means of all possible routes (contact, respiration, ingestion, root translocation, etc.).
- *Biomagnification:* this refers to the increase in concentration in living organisms by increasing the level in the food chains (e.g., level in predator > level in prey); biomagnification from one level to another is expressed by the *enrichment factor* (or *biomagnification factor*).

While bioaccumulation and bioconcentration phenomena may be influenced by species-specific biological factors, such as transport and distribution inside the organisms and metabolism, bioconcentration is essentially related to the physicochemical properties of the contaminants and can be described by a more simple approach, relatively independent of the species considered and of its ecological role. Bioconcentration can be thermodynamically quantified by means of the bioconcentration factor, BCF:

$$BCF = C_O / C_W \tag{1.75}$$

with C_O and C_W indicating the equilibrium concentrations in the organism and in water, respectively, in homogeneous units.

Bioconcentration was first studied in aquatic systems where the BCF, here also called K_{FW} or the fish/water partition coefficient, observed in laboratory models was tentatively correlated with K_{OW}. The first test (Neely et al., 1974), by means of eight selected chemicals and a fish species (the rainbow trout, *Salmo gairdneri*), led to the following relationship.

$$\log BCF = 0.542 \log K_{OW} + 0.124 \quad r^2 = 0.899 \quad n = 8 \tag{1.76}$$

Subsequently, Veith and collaborators (1979), using 55 different chemicals in different fish species (mainly fathead minnow, *Pimephales promelas*) obtained:

$$\log BCF = 0.85 \log K_{OW} - 0.7 \quad r^2 = 0.897 \quad n = 55 \tag{1.77}$$

Later, other investigators developed new correlations with similar conclusions, but with slope values (the coefficient of log K_{OW}) tending to reach unity, and intercept values (–0.7 in Equation 1.77) always negative and below –1. For example, Hawker and Connell (1986), with a daphnia and a marine mollusc, found, respectively:

$$\log BCF = 0.898 \log K_{OW} - 1.315 \quad r^2 = 0.925 \quad n = 22 \tag{1.78}$$

and

$$\log BCF = 0.844 \log K_{OW} - 1.235 \quad r^2 = 0.692 \quad n = 34 \tag{1.79}$$

All these are log/log correlations, where the slope 1 indicates that BCF and K_{OW} are of the same order and are directly proportional; the negative intercept between –1 and –2 means that, without logarithms, the proportionality coefficient is between 0.1 and 0.01.

At present, one of the most frequently used correlations to calculate BCF or K_{FW} is that proposed by Mackay (1982):

$$BCF = K_{FW} = 0.048\ K_{OW} \tag{1.80}$$

In this, the fish behaves as if it was 4.8% octanol (a value not so far from the average fish lipid content).

From the point of view of the dimensions, K_{FW} for fish originates from the concentration measurements in fish and water expressed in different units: mass/mass for fish and mass/vol for water. However, fish density is very near to 1 kg/L, and K_{FW} can be considered as it was expressed in (mass/vol)/(mass/vol) units, as for K_{OW} and K_{AW}. Correlations such as that in Equation 1.80 can be applied exclusively to nonpolar chemicals, where the bioconcentration follows passive mechanisms. For polar substances, active transport and reaction or complexation inside the organisms may lead to different results. However, the BCF approach can be used to compare direct measurements and calculations to investigate the possibility of particular bioconcentration processes.

In the case of "superhydrophobic" substances (log $K_{OW} > 6.5$), BCF values, calculated according to the previously discussed approach, are generally higher than measured (Figure 1.11). This may depend on several factors, such as the long time needed to approach equilibrium, or the differences existing in the solubility of superhydrophobic chemical in biological lipids and 1-octanol (Connell, 1990).

As far as bioconcentration is concerned, in terrestrial systems, there is something very similar to the pair fish/water in the aquatic ones: the pair leaf/air, particularly for hydrophobic substances, where root translocation and mobility

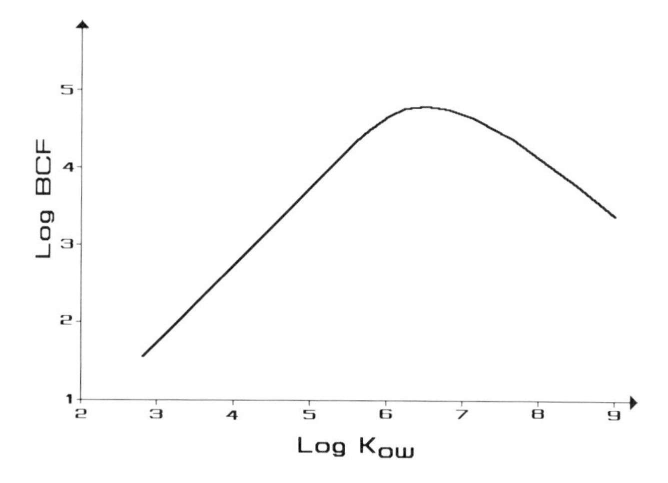

Figure 1.11. Relationship between log BCF and log KOW. (Modified from Connell, 1990.)

inside the plant are quite limited. Taking this observation as a starting point, some simple elaborations may help illustrate the analogy:

$$K_{LA} = C_L / C_A \tag{1.81}$$

with K_{LA} as the leaf/air bioconcentration factor, in (mass/vol)/(mass/vol) units; C_L equilibrium concentration in the leaf (mass/vol); and C_A equilibrium concentration in the air (mass/vol). As first suggested by Connell and Hawker (1986) and by Travis and Hattemer-Frey (1988), from the previous relationship:

$$K_{LA} = \left(C_L / C_W\right) / \left(C_A / C_W\right) \tag{1.82}$$

where C_W is the equilibrium concentration in water. When the system is saturated, C_W is the solubility in water (mass/vol); C_A the solubility in air, P/RT; and C_L the saturation concentration in leaf. But, similar to Equation 1.80:

$$C_L / C_W = L\, K_{OW} \tag{1.83}$$

where L indicates the lipid mass fraction in the leaf. Considering that $(P/RT)/C_W$ is K_{AW} (the air/water partition coefficient previously illustrated), from Equations 1.82 and 1.83 we have:

$$K_{LA} = L\, K_{OW} / K_{AW} \tag{1.84}$$

or:

$$K_{LA} = L\, K_{OA} \tag{1.85}$$

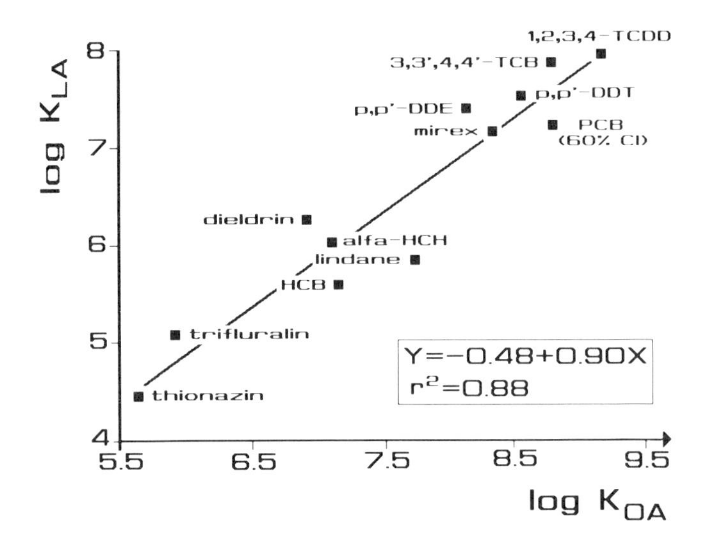

Figure 1.12. Correlation between the leaf/air bioconcentration factor KLA and the 1-octanol/air partition coefficient, K_{OA} in azalea *(Azalea indica)*. (Modified from Paterson et al., 1991. With permission.)

where $K_{OA} = K_{OW}/K_{AW}$, indicates the new 1-octanol/air partition coefficient introduced by Paterson et al. (1991). Results of a laboratory model with chemicals with a wide range of K_{OA} values and azalea *(Azalea indica)* leaves are reported in Figure 1.12, after selecting, for mirex, a K_{AW} value of 0.0322 (Yin and Hassett, 1986), as suggested by Bidleman (1991).

Considering the tendency to underestimate the bioconcentration factor for the more hydrophobic and less volatile chemicals, and the error intrinsic to the approach, to a first approximation and for practical purposes, the following correlation may be applied to calculate the leaf/air bioconcentration factor for hydrophobic and nonreactive substances (Bacci et al., 1990a,b):

$$K_{LA} = 0.022\ K_{OA} \tag{1.86}$$

where the proportionality constant 0.022 indicates the lipidic fraction of leaves.

Calamari and co-workers (1991) collected 300 samples of plants in 26 areas distributed worldwide and measured the levels of hexachlorobenzene (HCB), hexachlorocyclohexanes (HCHs), and DDT and related compounds (DDTs). They then calculated the ratio between median concentrations found in plant foliage and measured levels, taken from the literature, of the same chemicals in the air of corresponding areas. The average of these ratios, which indicate, an approximated leaf/air bioconcentration factor under different field conditions, has been compared with the corresponding bioconcentration factor from the azalea model. Results are shown in Table 1.6.

If the approach illustrated before seems to work quite well with noreactive

Table 1.6. Plant Foliage/Air Bioconcentration Factor K_{LA} in (mass/volume)/(mass/volume) Units: From the Field and From the Azalea Model for Some Nonreactive and Nonpolar Chemicals

	K_{LA}	
Chemical	**Field**	**Azalea model**
HCB	2.9×10^5	4.2×10^5
HCHs	2.5×10^6	1.0×10^6 (α-HCH)
		7.6×10^5 (γ-HCH)
DDTs	5.4×10^7	4.3×10^7 (*p,p′*-DDT)
		3.0×10^7 (*p,p′*-DDE)

Note: In the original paper, K_{LA} was expressed in (mass/mass)/(mass/mass) units; this last bioconcentration factor is 1/224 of the other one (Bacci et al., 1990a).

From Calamari et al., 1991. With permission.

chemicals, it may not be suitable when translocation and/or degradation phenomena occur: McCrady and Maggard (1993) have shown that the reed canarygrass *(Phalaris arundinacea)*, exposed to vapor of 2,3,7,8-tetrachlorodibenzo-*p*-dioxin, shows a BCF relatively near to that obtained with the azalea model, but only if the photodegradation of the chemical is not considered. After including photodegradation, which occurs for this substance under natural sunlight, the new BCF is reduced by about one order of magnitude.

1.3.4.3. Other Important Partition Coefficients

Mineral components of soils are able to retain polar chemicals, particularly when positively charged. A limitation of the K_p approach based exclusively on K_{OW} is the nonapplicability to the soil/water partition of polar compounds. In this case, a better approach is the experimental measurement (direct or from the literature). The soil/water partition coefficient for polar substances can be called K_D. This partition coefficient also has the same dimensional problems previously discussed for K_p. Thus, for polar chemicals, a dimensionless K_{DW} can be derived as follows:

$$K_{dW} = \rho_s K_D \tag{1.87}$$

where ρ_s is the soil density, (kg/L).

Another useful partition coefficient is that between aerosol particles, X, and air, A. This can be calculated as follows, as suggested by Mackay et al. (1986):

$$K_{XA} = 6 \times 10^6 / P_L \tag{1.88}$$

where K_{XA} is the aerosol/air partition coefficient under "clean" air conditions, and P_L is the vapor pressure of the chemical, (in pascal) as a subcooled liquid. This partition coefficient is of practical use when combined with another parameter: ϕ or the volume fraction of the particulates. This is typically 50×10^{-12} (m^3/m^3) in urban environments and 5×10^{-12} in rural environments; when $K_{XA}\ \phi = 1$, *equipartitioning between particulate and air occurs.* This is obtained when $6 \times 10^6\ \phi/P_L = 1$.

In urban environments:

$$6 \times 10^6 \times 50 \times 10^{-12} / P_L = 1 \tag{1.89}$$

consequently:

$$P_L = 3 \times 10^{-4}\ \text{Pa} \tag{1.90}$$

where P_L indicates the subcooled liquid vapor pressure value for equipartitioning in urban air.

By the same approach, it can be seen that, in rural environments, equipartition is expected when $P_L = 3 \times 10^{-5}$ Pa. Experimental measurements of the vapor-particle partitioning of organic chemicals (such as PCBs, PAHs, PCDDs, and PCDFs) led to results in accordance with this approach and constituted the basis for its development (Bidleman et al., 1986) and validation (Eitzer and Hites, 1989).

1.4. EVALUATIVE MODELS

Evaluative models provide an estimation of possible environmental trajectories of contaminants, indicating main reservoirs and sinks, and potential environmental targets. Although of limited applicability in predicting actual concentrations, they may be applied for preliminary analysis and in producing rankings.

1.4.1. Thermodynamic Models in Pairs of Environmental Phases

These represent the simplest approach: thermodynamic models are those not concerned with process rates, and thus, without the dimension of time: phase equilibrium and partition coefficients are in the domain of thermodynamics, while reaction, disappearance, and persistence in the environment concern kinetics. Despite their relative simplicity, thermodynamic evaluative models may be applied to assist the environmental chemist. Here are some examples of the application of equilibrium concepts to pairs of environmental phases.

Water and Fish. Often, in regulations for water and fish, allowed maximum concentrations may be conflicting: because of originating from different needs, it may happen that a given level in water is acceptable, while the corresponding level in fish is not acceptable. Let us suppose that we have field data on the concentration of a pesticide in lake water, C_W (mass/vol; e.g., mg/L). Data are not scattered

Table 1.7. Concentration in Fish, C_F, Measured and Predicted by the Fish/Water Bioconcentration Factor K_{FW} (based on K_{OW}) and Measured Levels in Water

Chemical	K_{OW}	K_{FW}	CW (mg/L)	CF calc. (mg/kg)	C_F measured (mg/kg)	Ref.
Atrazine	219[a]	10.5	0.001	0.01	0.01	d
TBT^+	5,000[b]	240	0.0005	0.12	0.9	e
CH_3Hg^+	1.6[c]	0.0768	<0.000001	<<0.000001	0.1	f

[a] Worthing and Hance, 1991.
[b] Major et al., 1991.
[c] Laughlin et al., 1986.
[d] Bacci et al., 1989.
[e] Unpublished results.
[f] Bacci, 1989.

and allow the calculation of the expected concentration of the pesticide in fish living in the lake. By the application of the bioconcentration factor K_{FW} level in fish, the C_F (in the same units of C_W) will be:

$$C_F = K_{FW}\,C_W \tag{1.91}$$

If K_{FW} is 1000 and C_W is 0.01 mg/L, the expected level in fish is 10 mg/L or 10 mg/kg, the fish density being close to 1 kg/L. To calculate K_{FW}, Equation 1.80 may be applied. In this case, the value of the bioconcentration factor is based on K_{OW}. This approach is quite appropriate for nonreactive and nonpolar chemicals: field measurements of C_F should not be far from the calculated value if the exposure time has been long enough.

With nonpolar reactive chemicals, it is expected that the measured C_F is lower than predicted by means of the K_{OW} approach, due to degradation of the chemical in the fish and the consequent reduction of the equilibration level. In the case of polar refractory chemicals, it may happen that conjugation with organic radicals may lead to levels in the fish that are higher than predicted, with a long time needed for equilibration, or even with an irreversible accumulation. In Table 1.7, a few examples are reported. In the case of the atrazine, the BCF approach seems correct, and measured data are in agreement with calculations. In the other two cases in Table 1.7, the underestimation of the levels in fish may be useful in indicating the presence of a combination of the simple diffusion (from water to fish), with some binding inside the organisms changing the chemical nature of the contaminant and contrasting or impeding release. In the case of methylmercury, it is well known that its binding to sulfhydryl groups of proteins leads to bioconcentration, bioaccumulation, and biomagnification in aquatic environments, despite its very low hydrofobicity (for methylmercury chloride, K_{OW} is about 0.3; Major et al., 1991).

Air and Foliage. The same approach as for fish and water can be applied to the pair leaf/air: from measured levels of an organic contaminant in plant foliage, by means of the K_{LA}, the corresponding concentration in the air can be obtained, and vice versa. Sometimes, there may be a considerable difference between calculated and measured data, as in the case of 2,3,7,8-tetrachlorodibenzo-*p*-dioxin discussed in Section 1.3.4.2. This does not necessarily imply that this kind of exercise is not useful; it simply means that the assumptions, or *boundary conditions,* of the model are incorrect and need to be modified. An example could better explain the usefulness of this approach, even when incorrect: the case of metallic mercury uptake, as vapor, in plant leaves.

The air/water partition coefficient, K_{AW}, of elemental mercury at 25°C is, according to Equation 1.69:

$$K_{AW} = \frac{P/(RT)}{S} = \frac{0.24/(8.314 \times 298)}{3 \times 10^{-4}} = 0.32 \qquad (1.92)$$

The vapor pressure P (in pascal) and water solubility S (in mol/m^3) are taken from Bodek et al. (1988). [R is the gas constant in Pa m^3/(mol·K) and T is in K.]

The octanol/water partition coefficient, K_{OW}, is 4.17 (Shoichi and Sokichi, 1985).

Given that mercury is not sufficiently hydrophobic, the leaf/air bioconcentration factor, K_{LA}, will be better obtained by the following relationship, introduced by Paterson et al., 1991:

$$K_{LA} = y_A + y_W K_{WA} + y_O K_{OA} \qquad (1.93)$$

where y_A is the air volumetric fraction in the leaf, y_W the water volumetric fraction and y_O the lipid volumetric fraction with values of the order of 0.2, 0.7, and 0.02, respectively; K_{WA} is the water/air partition coefficient, or the reciprocal of K_{AW} (K_{WA} = 1/K_{AW}; for mercury, K_{WA} = 1/0.32 = 3.125); the octanol/air partition coefficient is:

$$K_{OA} = K_{OW}/K_{AW} = 4.17/0.32 = 13 \qquad (1.94)$$

Thus, for elemental mercury:

$$K_{LA} = 0.2 + 0.7 \times 3.125 + 0.02 \times 13 = 2.6 \qquad (1.95)$$

K_{LA} being dimensionless, but originating from a ratio (mass/vol)/(mass/vol), this result means that a given volume of wet leaf (70% water content) should contain a quantity of mercury 2.6 times the quantity contained in the same volume of air, or that the concentration of mercury in the leaf, in mass/vol units, should be 2.6 times the concentration in air in the same units.

Taking a wet leaf density of 890 kg/m^3 and a background level of mercury vapor in the air of 3 ng/m^3, this leads to calculated levels in plant leaves of the order of 10 pg/kg (wet weight). This is very far from field data, currently indicating that, under background exposure to mercury vapors, plant foliage contains ~10 ng/g, or *one*

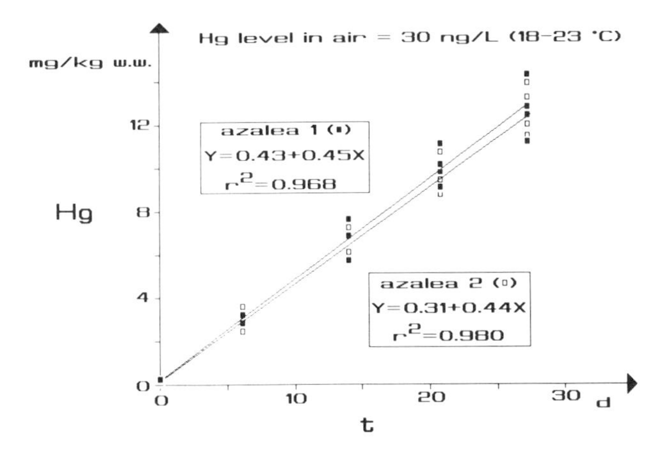

Figure 1.13. Irreversible accumulation of mercury vapors in azalea *(Azalea indica)* leaves. (Unpublished.)

million times the predicted concentration. This probably means that metallic mercury vapors, after reaching leaf cells, are rapidly transformed to mercuric ions; complexation may then occur producing new, less mobile substances which cannot be eliminated as easily as elemental mercury. The possibility of an irreversible accumulation of mercury in plant leaves, has been demonstrated by a laboratory model with azalea plants (Gaggi et al., 1991). Unpublished results obtained with similar experiments carried out in a semiclosed building, previously used for cinnabar ore roasting, with nearly constant concentration of mercury vapor in the air are shown in Figure 1.13; 1 week's sampling interval is enough to obtain good straight lines, indicating an irreversible accumulation. The reproducibility of the data, *each from a single leaf,* is shown. The mercury vapor level in the air was measured in 50-L samples by trapping mercury in 250-ml water bubblers (Drechsel gas washing bottles) containing 100 ml 1.5 N HNO_3 and 0.1 *M* $Na_2S_2O_8$, in water. This method, suitable for a high concentration of mercury in the air, taken from Braun et al. (1984) with slight modifications, was probably more accurate than the gold trap method applied in previous studies (Gaggi et al., 1991). Trapping efficiency was 75%, measured in three traps placed in series.

The slopes of the accumulation plots in Figure 1.13 are directly proportional to the mercury concentration in air. So, in this case, instead of the K_{LA}, the accumulation rates measured in plant leaves, normalized to a unitary concentration of mercury in the air, may be used to calculate the average mercury levels in air with an acceptable accuracy.

Another example of the usefulness of the K_{LA} approach, even if inappropriate, is the case of alachlor, a widely used herbicide. The K_{LA} obtained from experimental kinetics was 2.85×10^5, which is more than one order of magnitude below the value predicted by the correlation with K_{OA} (Bacci et al., 1990b). This data is in agreement with available literature, indicating a relatively high mobility and reactivity of alachlor in plant tissues: two output mechanisms not considered in the equilibrium partition coefficient K_{LA} for hydrophobic and conservative substances.

Sediments and Water. Information on the expected partition in this simple system can be extremely useful in correctly addressing research. In a meeting, a colleague said that the herbicide glyphosate did not contaminate surface waters: its application around a lake led to undetectable levels in lake waters after several storms. This is correct. Unfortunately, this chemical is strongly adsorbed by soils and practically all the glyphosate reaching the lake was in the sediment, not in water. The question was if the levels in sediment were or were not able to produce any adverse effect on the lake ecosystem.

The sediment/water partition coefficients K_P for hydrophobic compounds and K_D for polar chemicals may indicate the effectiveness of remedial or recovery actions: if the *reservoir* of the chemical is the water phase, an increase of the water turnover will shorten the time needed for restoration. *When the chemical is controlled by the sediment reservoir, even the open sea may not be adequate for a rapid recovery.* This is the case in a mercury polluted site, in the sea in front of a chloralkali plant eliminating inorganic mercury salts and carbonate particles. The ionic mercury is adsorbed by suspended particulate matter (organic, inorganic, living, or dead; $K_D > 100{,}000$) and settles, generating contaminated sediments, where slow methylation occurs producing bioavailable organic mercury able to promote bioconcentration, bioaccumulation, and biomagnification. This is less evident in terrestrial systems due to the oxic environment and to the weak resistance of methylmercury to photolysis (*cf.* Bacci, 1989), while in aquatic sediments with less oxygen available and no light, an excess of mercuric ions leads to an increase of net production of methylmercury, thus contaminating fish, particularly those feeding near the bottom.

In the open sea, a small artificial mercury deposit was generated by mercury-rich carbonates sedimentation (Renzoni et al., 1973). This contaminated sediment layer is now waiting for a complete burial (which is going on rapidly due to the carbonate emission; Bacci et al., 1984) by new carbonates which are, since the first half of the 1970s, practically mercury-free (background levels only). The halt of mercury contamination, of course, has been effective in reducing levels in fish; however, in 1991 a mean methylmercury concentration of 0.39 mg/kg fresh weight was still found in sea scorpions *(Scorphaena porcus)* of 25 to 50-g body weight (Table 1.8).

Air and Water. Air-water exchanges are often one of the key aspects in environmental distribution of organic chemicals. From a thermodynamic point of view, they are regulated by the air/water partition coefficient, K_{AW}. According to Mackay (1991), compounds with $K_{AW} > 10^{-1}$ are volatile and their transfer from water to air is controlled by their diffusion in water. The K_{AW} for water is 1.7×10^{-5} (20°C): taking into account that in natural conditions water vapor is already in the air, reducing the net flux from water to air of water, chemicals with a K_{AW} a couple of orders of magnitude below that of water ($K_{AW} < 10^{-7}$) are trapped in water.

Table 1.8. Methylmercury Concentrations in Muscle Tissue of Sea Scorpions *(Scorphaena porcus)* Caught Half a Mile from the Outfall of the Effluent of a Chloralkali Plant near Leghorn (Italy) in the period 1973–1991. Sampling Site: Secca de I Catini. Reference area: Torre Mozza (Vignale Riotorto; Leghorn)

Year	Methylmercury (mg/kg fresh weight)	Number of samples	Coefficient of variation (%)
Secca de I Catini			
1973	1.67[a]	20	12.5
1975	1.08[b]	21	18.8
1976	1.18[b]	5	12.6
1983	0.47[b]	14	31.6
1990	0.43[c]	12	25.7
1991	0.39[c]	19	16.6
Torre Mozza	0.05[b]	17	31.4

Note: Conversion of total mercury concentrations in methylmercury (years 1973–1983) by means of the correlation $Y = -0.078 + 1.02\ X$ (from Renzoni et al., 1991), where Y is the concentration of methylmercury (mg/kg) and X that of total mercury (in the same units).

[a] From Renzoni et al., 1973.
[b] From Bacci et al., 1984.
[c] Unpublished results.

Compounds with $10^{-5} < K_{AW} < 10^{-3}$ are relatively nonvolatile and air-phase diffusion controlled, but still able to give rise to significant water-to-air transfers. In a homologous series, the value of K_{AW} tends to be constant: substituting methyl groups or chlorines for hydrogen may reduce both vapor pressure and solubility by a factor 4 to 6 (Mackay, 1991). As a consequence, *less volatile compounds may reach the air from water if they are adequately hydrophobic.*

One practical consequence of K_{AW} is the difference in the role of respiration in uptake and clearance of hydrophobic chemicals in aquatic (i.e., gill-breathing) and terrestrial (lung-breathing) animals. Let us take a tuna fish, a dolphin, and a pair of organochlorine compounds—such as *p,p′*-DDE, with $K_{AW}=3.3 \times 10^{-3}$ and HCB, with $K_{AW} = 5.4 \times 10^{-2}$ (at 20°C). Jones et al. (1990) measured ventilation volumes in three species of tuna during swimming: kawakawa *(Euthynnus affinis),* yellowfin tuna *(Thunnus albacores),* and skipjack *(Katsuwonus pelamis),* finding average values ranging from 1.5 to 6.6 L/min per kg body weight or 90 to 400 L/(h·kg); a value of 100 L/(h·kg) can be selected for discussion. Dolphins, such as *Tursiops*, breathe two to three times per minute, less frequently than terrestrial animals, but with deeper and more effective (in extracting oxygen from the air) breaths. Tidal air ranging from 5.5. to 10 L was measured in *Tursiops truncatus* of 145 to 170 kg body weight (Irving et al., 1941) and maximum exhalations by 138-kg animals were 5 to 6 L (Ridgway et al., 1969); from these results, an air exchange rate of 10 L/(h·kg) can be selected for dolphins.

Assuming that the mechanisms of exchange (via respiration and via food intake) of nonpolar chemicals in dolphin and tuna have the same yield (in the following example, for simplicity, 100% is assumed), the relative significance of the exchange of these and other lipophilic compounds by respiration and by feeding can be evaluated (Figure 1.14). In order to perform calculations, the air/water partition of *p,p′*-DDE and HCB, the fish/water bioconcentration factor, and the food intake, estimated at 1 g/h per kg body weight for both dolphins and tuna fish, were taken into consideration.

Contact with a given volume of water implies contact with a mass of *p,p′*-DDE or HCB much greater than that of an equal volume of air. From the point of view of clearance, in the case of DDE, contact with that volume of water implies a potential for eliminating in excess of 10^3 times that of the dolphin. Consequently, the relative role of input via food and via respiration in these two animals greatly differ. *This could be the reason for the high biomagnification factor currently found for organochlorines in fish-eating mammals and birds as compared to fish (in the order of 100) and, at the same time, the reason for the relatively slight biomagnification occurring in predator fish.*

These considerations are not original: as a result of the pioneering work by Hunt and Bischoff (1960), biomagnification of the insecticide DDD (a DDT-related compound) was discovered in different food chain levels in a lake in California (Clear Lake), with the highest concentrations in a fish-eating bird (the western grebe, *Aechmophorus occidentalis*).

Soil and Air. In this pair of environmental phases, it is interesting and more useful to consider the volatilization potential rather than the soil/air partition. This can be evaluated by means of the simple approach introduced by Hartley (1969) for inert surfaces:

$$J = A\,P(M)^{1/2} \tag{1.96}$$

where A is a proportionality constant, J the vapor flux, P the vapor pressure, and *M* the molar mass of evaporating substance. In semiclosed systems with constant temperature and air turnover (e.g., a greenhouse), the previous relationship can be modified as follows:

$$C_A = B\,J \tag{1.97}$$

where C_A indicates the concentration in the air and B is a proportionality constant; combining the last two equations:

$$C_A = const\;P(M)^{1/2} \tag{1.98}$$

where *const* is a new proportionality constant, C_A is in mol/m^3, P is in Pa, and *M* is the molar mass, in g/mol. In small glass greenhouses (vol = 0.2 m^3), with an air turnover of 150 L/h, and sand contaminated with about 0.2 mmol of selected chemicals and kept moist, Bacci et al., (1992) found the following value for the constant: 10^{-7} mol/[(m^{3}Pa(g/mol)$^{1/2}$] at 20–25°C. Considering that the molar mass of selected chemicals was in the range 100 to 400 g/mol, the concentration in the

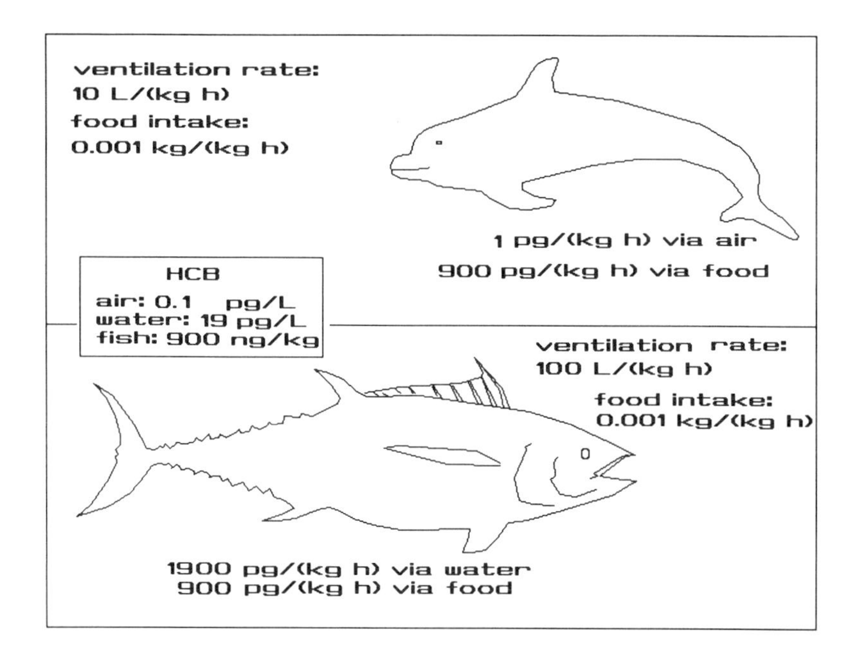

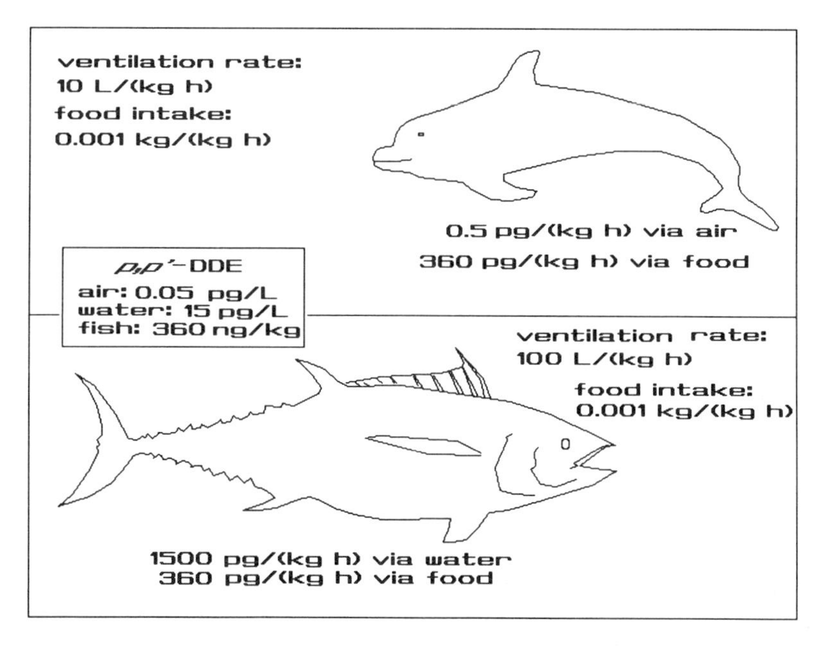

Figure 1.14. Relative role of intake via respiration and via food in air-breathing and water-breathing animals for hydrophobic semivolatile chemicals: the case of HCB (top) and *p,p'*-DDE (bottom).

air can be expressed, to a first approximation, as a function of vapor pressure and then as a fraction of the saturation concentration. In the experiments mentioned before, levels in the air were in the order of *0.5 to 1% of the chemical saturation concentration.*

For most of the pesticides, to a first approximation, the concentration in air over treated fields can be expected to be simply a constant fraction of the saturation concentration (Hartley and Graham-Bryce, 1980). For example, after an application of 1.5 kg/ha of nitrapyrin, Majewski et al. (1990) found an average concentration of 0.4 $\mu g/m^3$, or 1.1% saturation, at a height of 1.5 m. In another study, Majewski et al. (1993) during the first 3 days after the application of 2.5 kg/ha of trifluralin and triallate to fallow soil, found 1 m above the soil surface concentrations in air ranging from 0.39 to 10.60 $\mu g/m^3$ for trifluralin and from 0.48 (when trifluralin was 0.39) to 20.45 (and trifluralin 10.60) $\mu g/m^3$ for triallate (the first measurements, 28.4 and 35.9 mg/m^3 for trifluralin and triallate, respectively, were not considered). The high variation in field data was due to the variability of environmental conditions, such as the wind speed ranging from 1 to 10 m/s, air relative humidity (50 to 100%), and temperature (10 to 28°C). At the beginning of the second day, with a wind speed around 1 m/s and an air temperature of 13 to 18°C, levels of trifluralin (vapor pressure: 13.7 mPa, 25°C) and triallate (vapor pressure: 16 mPa, 25°C) were 6 and 8 $\mu g/m^3$, or 0.3 and 0.4% of the saturation concentration at 25°C.

Ross et al. (1990) collected air samples 30 m downwind from the edge of a circular 1-ha experimental onion field to investigate vapor losses and the possibility of reconcentration in plant leaves. At 1.5 m height from surface soil, treated with DCPA (chlorthal-dimethyl), they found DCPA vapor concentrations during the first 21 d after the treatment, ranging from 22 to 910 ng/m^3 with no relation with the time elapsed from treatment. The median value was about 100 ng/m^3. The field was regularly irrigated to enhance DCPA volatilization (Hartley and Graham Bryce, 1980; Spencer and Cliath, 1973). Mass balance 21 d after application indicated that about 10% DCPA had been dissipated by volatilization. The 0.5% saturation concentration of DCPA vapors in the air (20°C) is 225 ng/m^3. At a distance of 30 m from the treated field, a dilution is likely and these findings indicate that in the air over the field the concentration of DCPA was probably near 0.5 to 1% saturation.

The examples discussed above indicate that, under field conditions, there is great scattering of experimental measurements of pesticide vapor concentration due to uncontrolled temperature, wind speed and turbulence, water content in surface soil, and vegetation density. However, the *"0.5% saturation rule"* seems to work in evaluating median levels, with no relation if these are at ng, μg or mg/m^3. This simple approach, originally suggested by Hartley (1969) and Hartley and Graham-Bryce (1980), can be proposed for evaluative and comparative purposes.

To calculate the saturation concentration in air, or the *solubility in air,* the *Gas Law* may be applied:

$$PV = nRT \tag{1.99}$$

Table 1.9. Examples of Calculation of the 0.5% Saturation Concentration in Air, $C_{A0.5\%sat}$, for Some Pesticides (25°C)

Active ingredient	Vapor pressure P (Pa)	$C_{A0.5\%sat}$ (25°C) (mol/m³)
Dichlorvos	7000×10^{-3}[a]	1.41×10^{-5}
Butylate	100×10^{-3}[a]	2.02×10^{-7}
Aldicarb	13×10^{-3}[b]	2.62×10^{-8}
Cymoxanil	0.08×10^{-3}[b]	1.61×10^{-10}
Terbutryn	0.0004×10^{-3}[a]	8.07×10^{-13}
Cypermethrin	0.0002×10^{-3}[b]	4.03×10^{-13}

[a] Worthing and Hance, 1991.
[b] Suntio et al., 1988.

where P is the pressure in Pa, V the volume in m³, *n* the number of moles (mol), R the gas constant = 8.314 Pa m³/(mol·K), and T the absolute temperature (K).

From the Gas Law:

$$n / V = P / RT \tag{1.100}$$

If V = 1 m³ and P is the *vapor pressure, P/RT is the number of moles, n, corresponding to air saturation by the vapor of the chemical.* The 0.5% saturation concentration, $C_{A0.5\%sat}$, mol/m³, is then calculated as follows: $C_{A0.5\%sat} = 0.05\ P/(RT)$, where P is the vapor pressure (Pa), R the gas constant = 8.314 Pa m³/(mol·K), and T the absolute temperature (K). Examples of calculations of the 0.5% saturation concentration with different pesticides are shown in Table 1.9.

Concentrations such as those in Table 1.9 represent a simple *evaluative model* for ranking the potential to generate *vapor drift* from contaminated sites or agricultural fields. In the example, for solids (at 25°C), the vapor pressure of the solid state was used. This seems appropriate when the formation or presence of crystals (i.e., the solid state) is likely, as in the case of fields just treated with pesticides. In more general terms, with finely dispersed substances, the subcooled liquid vapor pressure values are more suitable. These can be calculated as previously indicated.

The vapor pressure of 10^{-6} Pa seems to be a cut-off value for volatilization from soils: for $P_L < 10^{-6}$ Pa, vapor movements from contaminated soils probably become negligible. The temperature dependance of vapor pressure means that volatilization from soil, on one hand, and adsorption onto solid particles and inclusion in water droplets, on the other, are related to environmental temperature. Cold climates will thus favor deposition, and warm climates will enhance volatilization.

Another remark concerns the possibility that an apparently nonvolatile chemical, if solid and with a high melting point, may become a substance that is relatively mobile in air when finely dispersed (subcooled liquid). In the case of refractory compounds, this soil-to-air transfer may be a key process in generating global contamination (Kurtz, 1990; Calamari et al., 1991).

1.4.2. Thermodynamic Multimedia Models: The Fugacity Approach

The environmental fate of contaminants is determined by the joint action of several factors ascribable either to the nature of the substance or to the environment. So, on one hand, the physical and chemical properties will define the potential for reactivity and mobility and, on the other, the environmental variables will control the degree to which these potentials operate. Under field conditions, environmental variables (e.g., temperature, pH, light wavelength and irradiation intensity, water and air turnover and turbulence, living organisms, etc.) are quite complex to analyze and may produce significant changes in the environmental behavior of contaminants. This makes the physicomathematical models for environmental chemistry very complex and difficult to apply, especially when several details are considered with the aim of producing a simulation of a *real system.*

An alternative approach consists of developing simple models simulating *evaluative environments* in which the environmental variables are standardized and reduced to the essential *(evaluative models;* Baughman and Lassiter, 1978*).* In this case, the different behavior of various chemicals will essentially depend only on their physicochemical properties *(intrinsic properties).* The main limitation of evaluative models is that the results are not directly related to real situations and, consequently, the absolute data (e.g., an expected concentration in a given environmental compartment) cannot be transferred to the field. However, the results of a series of chemicals may be applied, for instance, to *rank* the potential of different substances to reach and contaminate water bodies or to rank their potential to generate vapor drift from the site of release. Data from evaluative models may also be *calibrated* by means of laboratory and field measurements of selected reference chemicals to produce information on the order of magnitude of the expected environmental levels.

The simplest multimedia evaluative models are those based on equilibrium among all environmental phases. These are assumed to be *homogeneous*, with a defined *accessible volume,* where the chemical is free to move from one to another, until equilibrium is reached. In these models, the variable *time* is not considered (thermodynamic systems).

These models are useful in getting a general picture of major trends in chemical partition. No reaction is considered nor time needed to reach the final equilibrium. Despite their intrinsic limitations, in some particular cases, thermodynamic models may be able to describe real situations.

The most illuminating approach to these kind of models was that introduced by Mackay (1979) and Mackay and Paterson (1981), where *fugacity* was applied as the driving force for partition. Fugacity has the dimensions of a pressure, its current units are pascals, and it expresses the *escaping tendency* of a substance from a phase (e.g., water, air, or soil). This concept is more easily understood than entropy or chemical potential, and it has a clear physical meaning since 1 molecule exerts 1/2 the escaping tendency of 2. Equilibrium among n phases is reached when all escaping tendencies are equal. In a phase, fugacity, f (Pa), is directly proportional to the concentration in that phase, C (mol/m^3) or:

$$C = Zf \tag{1.101}$$

where the proportionality constant Z is the *fugacity capacity,* dimensionally a mass/(vol. × pressure), with units of mol/(m^3·Pa). According to Mackay (1991), the Equation 1.101 does not necessarily imply that C and f are always linearly related. Nonlinearity can be included by varying Z as a function of C or f. The fugacity capacity Z will depend on the intrinsic properties of the chemical and on environmental variables, such as the nature of the medium and temperature of the system; at low concentrations, concentration effects are negligible.

In a two-phase system, at equilibrium we have $f_1 = f_2$ and, consequently, $C_1/Z_1 = C_2/Z_2$ or:

$$C_1 / C_2 = Z_1 / Z_2 = K_{12} \tag{1.102}$$

where K_{12} is the dimensionless partition coefficient between phase 1 and phase 2 and corresponds to the ratio of the respective fugacity capacities. This means that *each Z value is half a partition coefficient.* It *indicates the capacity of a phase for a chemical* and significantly simplifies the work in the case of a system with several compartments. In a system with 10 different phases, there are 90 possible partition coefficients, each related to each other, with only 9 of them independent, while by means of 10 Z values all possible partition coefficients can be obtained.

The Z value for air, Z_A, is:

$$Z_A = 1 / RT \tag{1.103}$$

at 20°C: $Z_A = 4.1 \times 10^{-4}$ mol/(m^3·Pa); at saturation f = vapor pressure (P) and the concentration in air, C_{Asat}, becomes $C_{Asat} = P/RT$.

For water,

$$Z_W = Z_A \, 1 / K_{AW} = S / P \tag{1.104}$$

where S and P are solubility in water (mol/m^3) and vapor pressure (Pa), both referring to the liquid state or, where appropriate, to the solid state ($Z_W = S_L/P_L = S_C/P_C$).

For soils and sediments,

$$Z_S = \rho_s \, K_p /(P / S) = K_{PW} /(P / S) \tag{1.105}$$

where ρ_s is the density (kg/L), P and S as before, and K_p the soil/water partition coefficient for undissociated substances.

Similarly, a Z_F can be defined for fish as representative of aquatic organisms:

$$Z_F = K_{FW} /(P / S) \tag{1.106}$$

where K_{FW} is the fish/water bioconcentration factor, as previously defined.

For plant foliage, a Z_L may be calculated:

$$Z_L = K_{LA}\ 1/(RT) \tag{1.107}$$

Taking $K_{LA} = 0.02\ K_{OW}/K_{AW}$ (Bacci et al., 1990b):

$$Z_L = 0.02\ K_{OW}/(K_{AW}\ RT) = 0.02\ K_{OW}/(P/S) \tag{1.108}$$

For aerosol, Z_X:

$$Z_X = K_{XA} Z_A = 6 \times 10^6 /(P_L\ RT) \tag{1.109}$$

where K_{XA} is the aerosol/air partition coefficient calculated according to Mackay et al. (1986), and P_L indicates the subcooled liquid vapor pressure.

Once the chemical, its properties, and load to the system have been established and the environmental compartments defined with relative accessible volumes, a simulation of the equilibrium distribution may be obtained as illustrated in the following example.

Fugacity calculations, Level I, partition of a hypothetical chemical A in a multimedia system composed of air, water, and soil:

Compartments: air, water, soil
Accessible volumes (m^3): $V_A = 10^6$; $V_W = 10^4$; $V_S = 100$
Phase densities (kg/L): air = 0.0012
water = 1
soil = 1.5
Organic carbon in soils, mass fraction: 0.02
Chemical: A, load: 100 mol
Melting point: liquid
Working temperature: 20°C, 293 K → RT = 8.314 × 293 = 2436
Molar mass: 200 g/mol
Solubility in water, S: 1 mol/m^3
Vapor pressure, P: 0.1 Pa
$\log K_{OW} = 4$; $K_{OW} = 10{,}000$

Z values [mol/(m^3·Pa)]:

$$Z_A = 1/RT = 4.1 \times 10^{-4}$$

$$Z_W = S/P = 10$$

$$Z_S = K_{PW}/(P/S) = \rho\ K_{OC}\ f_{OC}/(P/S) = \rho\ 0.41\ K_{OW}\ 0.02/(1/10) = 1230$$

The total quantity, Q (mol), of the chemical in the system is:

$$Q_T = \Sigma C_i V_i = f \Sigma V_i Z_i$$

where C_i and V_i indicate the concentration (mol/m^3) of the chemical in each phase and the volume (m^3) of each phase (air, water, and soil). At equilibrium (which makes f the same in all phases):

$$f = Q / \Sigma V_i Z_i$$

Q is known: in this example 100 mol; V is defined for each compartment; and Z values have been calculated. So the fugacity f (Pa) is obtained:

$$f = 100 / \left(V_A Z_A + V_W Z_W + V_S Z_S\right)$$

$$f = 100 / \left(10^6 \times 4.1 \times 10^{-4} + 10^4 \times 10 + 100 \times 1230\right) = 4.48 \times 10^{-4} \text{ Pa}$$

To avoid totally unrealistic simulations a *check for saturation* is essential. *If the system is saturated, the fugacity from the air will be equal to the vapor pressure of the chemical.* In the example, with fugacity being more than two orders of magnitude below the vapor pressure, the system is not saturated.

The quantity of chemical in each compartment, Q_i (mol) can now be calculated:

$$Q_A = C_A V_A = Z_A V_A f = 4.1 \times 10^{-4} \times 10^6 \times 4.48 \times 10^{-4} = 0.18 \text{ mol}$$

or 0.18% of the load, Q.
For water:

$$Q_W = C_W V_W =_W V_W f = 10 \times 10^4 \times 4.48 \times 10^{-4} = 44.8 \text{ mol, } 44.8\% \text{ of Q}$$

For soil:

$$Q_S = C_S Z_S = Z_S V_s f = 1230 \times 100 \times 4.48 \times 10^{-4} = 55.104 \text{ mol,}$$

or 55.1% of the load.

The sum of Q_A, Q_W, and Q_S is a little different from Q, due to approximation of decimals.

From Q values and from volumes and densities of the environmental compartments, the concentrations in mass/mass or mass/vol units can easily be obtained. Fugacity calculations are currently carried out by means of available software (Mackay, 1991), particularly helpful in the case of more complicated levels.

Chemicals with different properties will show different potential distribution patterns: hydrophobic semivolatile chemicals (e.g., DDT) will tend to partition in soil and sediment phases, while more water soluble and less volatile substances (e.g., atrazine) will tend to be mainly distributed in water. Compounds such as the chlorofluorocarbons (CFCs), responsible for the stratospheric ozone destruction, will partition in the air.

1.4.3. Kinetic Models

In the previous section, all selected evaluative models were thermodynamic (without the *time* as a variable). The introduction of time and velocities may add some complications. The lowest level of complexity is when the system is in *equilibrium* (all fugacities are equal) and in *steady state* (input = output). This corresponds to Level II in fugacity models by Mackay (1991).

The main difference between Level I and Level II products is that, with the former, only potential *reservoirs* are pointed out; while with the latter, potential environmental *sinks* can be seen, and the respective role of chemical reaction and advection, in determining environmental persistence, evaluated. Thus, opening the system to air and/or water inflow and outflow and introducing chemical transformation processes, but with the limitations due to equilibrium and steady state, we are still far from real field conditions. However, the model may supply useful information when appropriate time and space scales are selected.

To introduce reaction and advection loss mechanisms, first-order kinetics is assumed. In a given phase, the overall reaction rate may be expressed as:

$$N_R = k_R \, CV \tag{1.110}$$

N_R indicating the overall reaction rate in the considered phase (mol/h) k_R the overall reaction rate constant (h^{-1}), C the concentration of the chemical in the phase (mol/m^3), and V the phase volume (m^3). In as much as $C = Zf$:

$$N_R = k_R \, VZf \tag{1.111}$$

If $D_R = k_R$ VZ, then

$$N_R = D_R f \tag{1.112}$$

where D_R is a *transport parameter,* in units of mol/(Pa·h). Multiplied by fugacity, D_R values give the reaction rates (they are analogous to rate constants which, multiplied by a concentration, give the rates).

In the case of advection, by a similar approach we have:

$$N_A = k_A \, VZf \tag{1.113}$$

k_A being the advection rate constant (h^{-1}) previously defined. Thus, for advection, we also have:

$$N_A = D_A f \tag{1.114}$$

with

$$D_A = k_A VZ \tag{1.115}$$

D_A is again a transport parameter in units of mol/(Pa·h). Multiplied by fugacity, D_A values give the advection rates. Fast processes are characterized by large D values.

In a defined Level II evaluative system, all D values can be obtained, the advection and reaction rate constant being known (or better, required by the simulation), volumes of the various phases defined and Z values calculated as for Level I.

Rewriting the fugacity equation, for each phase:

$$f = I/(D_R + D_A) \tag{1.116}$$

with I = input rate (mol/h). Therefore, if D_R is 0.2 mol/(Pa·h), D_A is 0.3 mol/(Pa·h), and the input is 10 mol/h, the fugacity will be f = 10/(0.2 + 0.3) = 20 Pa.

To calculate fugacity in a Level II model, two kinds of information are needed:

- Input I (= output, due to the steady-state assumption)
- D values (previously illustrated)

To the forcing equation describing the input, a quantity of contaminant directly introduced into the system can be added, with the limitation that this quantity has to reach the system at a constant input rate E (mol/h); the input of contaminant due to mass transfer [an inflow of water at G m^3/h, containing a concentration C (mol/m^3), of the chemical] may also be considered:

$$I = E + \Sigma GC \tag{1.117}$$

The output O (mol/h), in a steady-state condition, can be expressed:

$$O = I = E + \Sigma GC \tag{1.118}$$

However, the output from the system is also:

$$O = \Sigma D_A f + \Sigma D_R f = \Sigma(D_A + D_R) f \tag{1.119}$$

which expresses the output as the sum of all reaction and advection processes. D values are homogeneous and can be added (as for first-order rate constants). From Equations 1.118 and 1.119:

$$f = (E + \Sigma GC) / \Sigma (D_A + D_R) \qquad (1.120]$$

or

$$f = I / \Sigma D \qquad (1.121)$$

Now, for a given system, all values for determining I and each D value are available: fugacity can be calculated. *It is important to observe that in Level II models, the input is defined, but not the quantity of contaminant in the system.* This quantity Q is important for further consideration and needs to be calculated:

$$Q = \Sigma Q_i \qquad (1.122)$$

where Q_i are the quantities in each phase, at equilibrium. These can be obtained as for Level I:

$$Q_i = C_i V_i = f Z_i V_i \qquad (1.123)$$

and Q will be:

$$Q = f \Sigma Z_i V_i \qquad (1.124)$$

The ratio I/Q will indicate the first-order disappearance constant k_{dis} (h^{-1}), containing information on joint transport and transformation mechanisms. Its reciprocal will be the overall persistence of the chemical in the system, or weighted mean residence time or turnover time T (h):

$$T = Q / I \qquad (1.125)$$

The Level II calculations will also indicate, together with persistence, the relative weight of loss mechanisms (advection and reaction), as well where the main environmental sinks are expected, (i.e., the compartment(s) where the reaction or degradation of the chemical may occur to the greatest extent).

An example of Level II fugacity calculations applied to Tributyltin (TBT) in a harbor area.

During the 1960s, triorganotin compounds, and particularly tributyltin (TBT), began to be used in marine antifouling treatments to protect cooling-water pipes of coastal power and industrial plants, as well as in antifouling paints for boats, ships, and other marine structures. The rapid rise of the use of TBT was due to its very high biocidal activity against fouling organisms, coupled with a relatively low toxicity to mammals. The high toxicity of TBT to nontarget species caused growing concern during the 1970s, particularly after the disasters of the French oyster industry, based on the Pacific oyster *(Crassostrea gigas)*: oysters exposed to relatively low levels of TBT, released by marine antifouling paints, suffered reproductive failure, severe shell thickening, and malformations (Alzieu, 1986).

Within the framework of the Mediterranean Action Plan of UNEP (United Nations Environment Programme), a pilot survey was started in 1988 to evaluate the status of contamination by TBT in harbors and marinas in selected sites along the Mediterranean coast (Gabrielides et al., 1990; Alzieu et al., 1991). One of these sites was the Leghorn harbor area (Figure 1.15) where concentrations of TBT in harbor waters were of the order of 100 to 1000 ng/L, and ranged from 400 to 810 ng/L in sites representative of the main water mass (Bacci and Gaggi, 1989). The TBT input into the system was of the order of 13.5 kg/d, with 10 kg/d originating from the antifouling treatment of a cooling pipe from a thermoelectric power plant and the remaining from the paint of pleasure boats and large ships. Water turnover time in the *antiporto* area (Figure 1.14: sampling sites 8, 9, 13) is of the order of 1 d. Field measurements indicated that the system was not far from a steady state, concentrations of TBT in the water being, within a factor of two, constant in different period of the years 1988–1989 (Bacci and Gaggi, 1990). This chemical is rapidly degraded to inorganic tin by the progressive removal of the organic groups linked to the tin atom. The application of a thermodynamic partition model is therefore inappropriate; besides, water turnover is large and should play a major role in TBT removal from the harbor. If degradation rates in sediments and water are known, a steady-state fugacity Level II model with advection can be applied.

The chemical is taken to be TBT chloride (TBT-Cl), the major species in marine environments (Laughlin et al., 1986). The system is open; only water and sediment are considered, due to the negligible volatility of TBT-Cl from water. Water and sediment concentrations of TBT-Cl are assumed to be at equilibrium. A first-order degradation rate constant of 0.00413 h^{-1} (corresponding to a half-life of 7 d) is applied to the water compartment (Lee et al., 1987); for sediment, a first-order degradation rate constant of 0.00018 h^{-1} (half-life of 160 d; Stang and Selingman, 1986) can be taken.

The physical properties of TBT-Cl are as follows:

Molar mass: 325.49 g/mol
Melting point = liquid
K_{OW} = 5000 (from: Laughlin et al., 1986)
Vapor pressure, $P_L < 1 \times 10^{-6}$ Pa; taken to be 1×10^{-6} Pa
Solubility in water, $S_L = 3 \times 10^{-5}$ mol/m^3 (from Blunden et al., 1984)

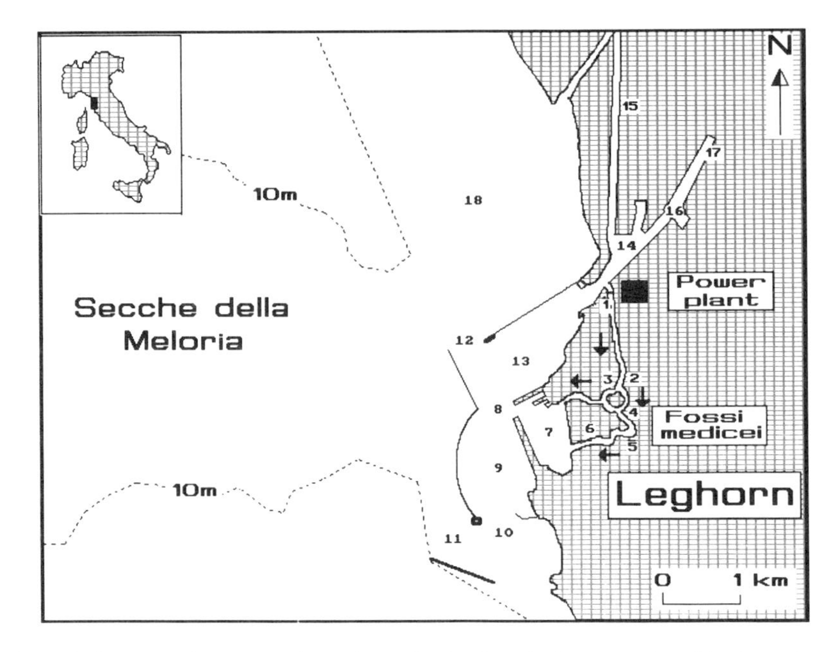

Figure 1.15. Location of sampling sites in the Leghorn harbor area. (Modified from Bacci and Gaggi, 1989.)

Properties of the system:

Water volume, $V_W = 2.7 \times 10^7$ m^3 (surface area = 3×10^6 m^2, 9 m average depth)
Water inflow, $I_W = 1.125 \times 10^6$ m^3/h
Water outflow, $O_W = 1.125 \times 10^6$ m^3/h
Sediment volume, $V_S = 1.5 \times 10^5$ m^3 (area as for water; depth 0.05 m)
Sediment density, $\rho = 1.5$ kg/L
Organic carbon, mass fraction = 0.02
Input rate for TBT-Cl into the system, E = 1.728 mol/h (constant)

Reaction kinetics:

Degradation in water: first-order rate constant, $k_{degW} = 4.13 \times 10^{-3}$ h^{-1}
Degradation in sediment: first-order rate constant, $k_{degS} = 1.8 \times 10^{-4}$ h^{-1}

Advection:

Water advection rate constant, $k_A = I_W/V_W = 4.17 \times 10^{-2}$ h^{-1}

Calculating TBT-Cl input rate, I in mol/h:

I = E = 1.728 mol/h

Calculating D values:

For reaction in water, $D_{RW} = k_{degW} V_W Z_W = 3{,}345{,}942$ mol/(Pa·h)
For advection in water, $D_{AW} = k_A V_W Z_W = 33{,}750{,}000$ mol/(Pa·h)
For reaction in sediment, $D_{RS} = k_{degS} V_S Z_S = 49{,}851$ mol/(Pa·h)

Calculating fugacity, *f*:
$f = I/\Sigma D_i = 4.65 \times 10^{-8}$ Pa

Calculating TBT-Cl concentrations and partition into the system:
Concentration in water $C_W = Z_W f = 1.39 \times 10^{-6}$ mol/m^3 = 452 ng/L
Quantity of TBT-Cl in the water phase, Q_{TBTW}:
$Q_{TBTW} = C_W V_W = 37.5$ mol (12.2 kg)
Concentration in sediment $C_S = Z_S f = 8.58 \times 10^{-5}$ mol/m^3 =
5.72×10^{-5} mol/t = 18.6 ng/g
Quantity of TBT-Cl in the sediment phase, Q_{TBTS}:
$Q_{TBTS} = C_S V_S = 12.9$ mol (4.2 kg)

The total quantity Q_{TOT} of TBT-Cl at equilibrium will be about 50 mol, 25% in the sediment and 75% in the water.

Persistence, T, of TBT-Cl in this system appears to be very short:
$T = Q_{TOT}/I = 50/1.728 = 29$ h

The total reaction rate is subdivided as follows:
Water: $D_{RW} f = 0.156$ mol/h
Sediment: $D_{RS} f = 0.0023$ mol/h

Advection:
Water: $D_{AW} f = 1.57$ mol/h

The reaction will occur mainly in the water phase, which is also the main reservoir of the chemical; advection contributes to more than 90% of the chemical output from the system (Figure 1.16).

The concentration of TBT in the water resulting from the model is similar to the field data. Of course, this does not necessarily mean that the model perfectly reflects a real situation. However, main trends in partition and the relative weight of reaction and advection probably are not far from reality. Furthermore, the model may be essential in indicating research needs such as the measurement of:

- the water/sediment partition coefficient
- degradation rates in water and sediment
- contamination levels outside the harbor due to the advection of TBT

One of the critical points of the model may seem to be the high water turnover selected (turnover time = 1 d). It could be said that the significance of TBT advection by water is only due to this high turnover. Analyzing the sensitivity of the model to this parameter (water inflow), it can easily be seen that if water turnover is reduced by one order of magnitude (i.e., from 1.125×10^6 to 0.1125×10^6 m^3/h, corresponding to a turnover time of 10 d), the persistence of TBT in the system will increase to 160 h, concentrations in water and sediments will rise by a factor of 5, and disappearance from the system will be negligible via

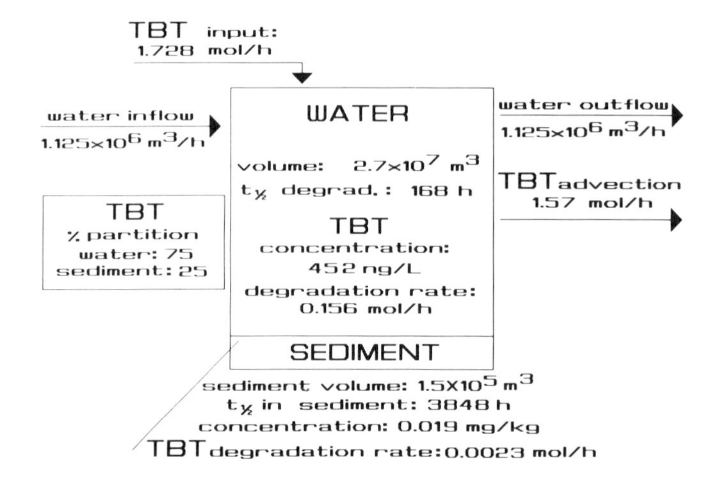

Figure 1.16. A Level II fugacity model for TBT in the Leghorn harbor area (steady state and equilibrium).

degradation in sediments (0.7% of the total), reaching 49.4 and 49.9% of the total with, respectively, degradation in water and advection with water renewal. Even under these conditions, water advection is still the main way out for TBT.

The fugacity approach introduced by Mackay and Paterson is continuously refined and it is suitable for simulating more complex systems (nonequilibrium and nonsteady state), but these are beyond the objective of this book.

1.4.4. Ranking the *Contamination Potential*

A possible application of evaluative models is for ranking a group of chemicals in relation to a defined environmental impairment. A relatively new problem arising from the application of pesticides in agriculture is that of *inadvertent residues in crops where they are not allowed.* Another priority topic in environmental chemistry is the protection of the quality of water reserves. In the following sections, two new tools are proposed as a call for research to improve present knowledge in environmental chemistry.

1.4.4.1. Vapor Drift and Inadvertent Residues *in Crops*

Vapor drift concerns the displacement of a chemical from one site to another by volatilization and transport via air. The air, in the field, is the most mobile environmental phase and, *though concentrations attained are low, the fluxes may be important,* due to high turbulence and advection. In general, air is the typical environmental compartment to which the famous phrase *dilution is solution to pollution* may be applied. However, the presence of low concentrations may hide significant mass transfer from one phase to another, particularly when reconcentration phenomena occur.

Vapor movements from treated fields may generate air concentrations of the order of 0.5% of saturation values and plant tissues (mainly "green" tissues; i.e., leaves) are able to take up pesticide vapors as a function of their leaf/air partition coefficient. Considering that present regulations in force in several countries do not allow commercialization of crops containing detectable amounts of pesticides not registered for use with that crop (with only a few exceptions for known soil carry-over effects in rotational crops and for global contaminants, such as certain organochlorinated pesticides), the presence of a pesticide in a nontarget species may lead to economic damage to farmers. This possibility is not so terribly remote: as reported by Ross et al. (1990), the California Department of Food and Agriculture found detectable residues of the herbicide 2,3,5,6-tetrachloroterephthalate, DCPA, on a variety of produce samples from Monterey County and from the Central Valley in San Joaquin County to which it had not been applied. The herbicide was found in more than 10% of the analyzed samples of daikon (*Raphanus sativus* L.), dill (*Anethum graveolens* L.), kohlrabi (*Brassica oleracea* L.), and parsley (*Petroselinum crispum* Mill.) at levels of the order of 10 to 100 ng/g wet weight. To find an explanation for these facts, Ross et al. (1990) carried out the following experiment: parsley plants were grown in pots around a circular field (1 ha) planted with onions (*Allium cepa* L.) and treated with DCPA at a rate of 7.08 kg/ha (DCPA is a herbicide currently applied to onion crops). Residues in the onions were well below the tolerance level (1 mg/kg, Code of Fed. Reg., 1988) and parsley planted back in the same soil, after the onion harvest, did not contain detectable levels of DCPA (detection limit = 20 ng/g) when sampled 217 and 336 d after the last application. However, the plants in pots installed around the onion field after the treatment (to avoid particulate drift) contained residue concentrations ranging from 58 to 640 ng/g, 23 m from the edge of the onion field, 10 d after the DCPA application.

Wind speed ranged from 1 to 6.9 m/s and temperatures were as follows: average maximum 29°C and average minimum 9°C; average relative humidity ranged from 85 to 21%. Air samples, collected 30 m downwind from the edge of the circular onion field at a height of 1.5 m, showed DCPA vapor concentrations during the first 21 d after the treatment, ranging from 22 to 910 ng/m^3, with no relation with the time elapsed from treatment. The median value of DCPA vapor concentration in the air was about 100 ng/m^3. The field was regularly irrigated to enhance DCPA volatilization (Spencer and Cliath, 1973). Mass balance 21 d after the application indicated that about 10% of DCPA was dissipated by volatilization (Figure 1.17).

The 0.5% saturation concentration of DCPA vapors in the air (20°C) is 225 ng/m^3 (vapor pressure= 3.3×10^{-4} Pa; DePablo, 1981). At a distance of 30 m from the treated field, a dilution is likely but, due to the approximation of the proposed approach, it is not taken into consideration. The concentration of 225 ng/m^3 is not far from the measured values. Levels in parsley exposed to a DCPA concentration in air of 225 ng/m^3 can be calculated by the leaf/air bioconcentration factor, K_{LA}, developed by Bacci et al. (1990b):

$$K_{LA} = 0.022\ K_{OW} / K_{AW}$$

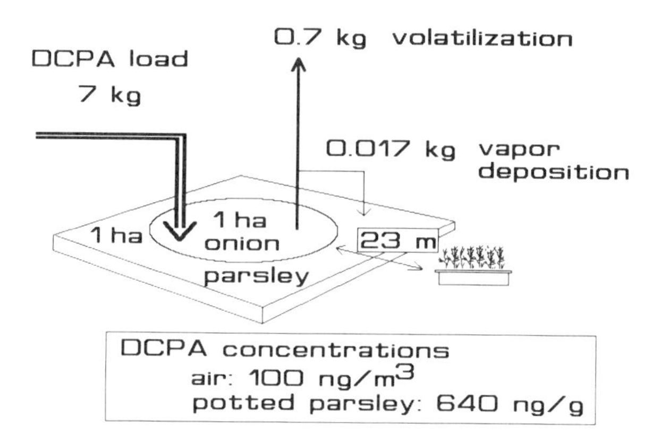

Figure 1.17. Mass balance 21 days after application of DCPA. Concentrations in air 30 m downwind; potted parsley plants 23 m downwind, time of exposure: 10 days. (Data from Ross et al., 1990.)

where K_{OW} and K_{AW} are the 1-octanol/water and the air/water partition coefficients, respectively (K_{AW} is the ratio of the solubilities in air and water, in mass/volume units). In Figure 1.18, calculations based on a combination of the 0.5% approach and K_{LA} are shown.

The data of Figure 1.18 are in good agreement with field measurements. The concentration of DCPA in parsley is very near to the measured value. The approximations intrinsic to the applied approach, however, indicate that the good correspondence found in this case cannot be taken as a rule. These findings therefore represent a call for research, rather than a well-established approach to be applied in the field. A combination of the *0.5% approach* and the leaf/air bioconcentration factor, K_{LA}, however may be tentatively applied to evaluate the potential of pesticides to volatilize from soils and concentrate in plant foliage. This rough and simple estimation method could be used as a starting point to rank the potential of different substances to generate *inadvertent residues* in crops by vapor drift. In principle, the approach is not suitable for chemicals which are higly reactive and/or mobile inside the plant. However, it can be applied, as a screening tool, to identify those active ingredients for which further studies are needed.

1.4.4.2. Ranking the Contamination Potential for Water Compartments

Different chemicals, including pesticides, after release into the environment, will move and react according to their "vocation" and to the environmental variables. If we assume that the relative behavior will be similar under different environmental conditions, a ranking of the potential to reach water compartments can be produced, essentially based solely on the intrinsic properties of the chemicals and on standardized environmental properties. The "number" needed to produce these ranks is called the *leaching index*. Several leaching indices have

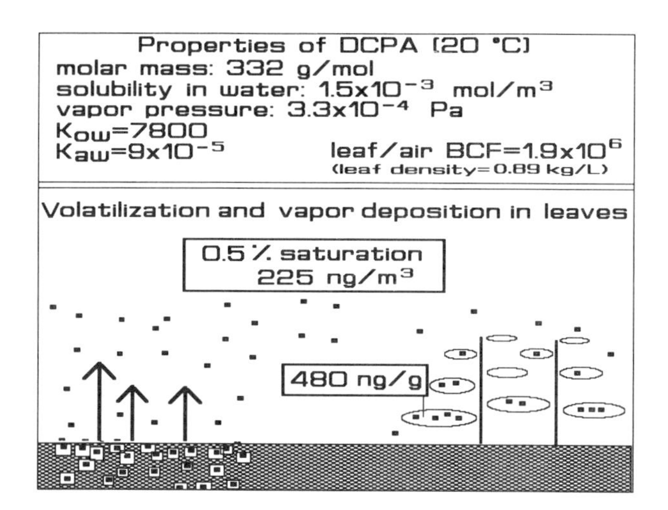

Figure 1.18. Evaluative model for off-target deposition of DCPA vapors on parsley.

been proposed (Wagenet and Rao, 1990; Trevisan et al., 1991). The first one is the *"convection time"*, t_c, developed by Jury and co-workers (Jury et al., 1983; Jury et al., 1984). The index is based on chromatography principles and on the assumption of an equilibrium distribution of the chemical between liquid, vapor, and adsorbed phases:

$$t_c = L / V_E = L\left(\rho_S K_D + FC + aK_{AW}\right) / J_W \qquad (1.126)$$

where t_c is in days (d), L is the thickness of the soil layer (m), V_E is the convective velocity (m/d), ρ_s is the soil bulk density (kg/L), K_D is the soil/water partition coefficient (L/kg), FC is the volumetric fraction of water content of the soil (field capacity), a is the volumetric fraction of air content of soil, K_{AW} is the air/water partition coefficient in (mass/volume)/(mass/volume) units, and J_W is the water flux or net recharge rate (m/d). The index may be applied to evaluate the relative mobility of a series of chemicals, calculating the time needed to cross a given soil layer, but does not take chemical transformation processes into account.

A similar approach was proposed by Rao et al. (1985), in which the *convection time*, t_c, was called *travel time*, t_r, and the focus was on a *retardation factor,* RF, (Davidson et al., 1968) affecting the *travel time* as indicated by rearranging Equation 1.126:

$$t_r = L\,\mathrm{RF}\,\mathrm{FC} / J_W \qquad (1.127)$$

where

$$\mathrm{RF} = 1 + \rho_S K_D / \mathrm{FC} + aK_{AW} / \mathrm{FC} \qquad (1.128)$$

Since the application of the retardation factor was restricted to nonpolar chemicals, the soil partition coefficient K_D was replaced by the product f_{oc} K_{oc}, where f_{oc} is the organic carbon fraction (mass/mass) and K_{oc} the organic carbon/water partition coefficient (L/kg):

$$RF = 1 + \rho_S F_{oc} K_{oc} / FC + a K_{AW} / FC \tag{1.129}$$

In Equation 1.127, the component L FC/J_W of the travel time indicates the travel time for water: water (RF = 1) takes 50 d to cross a soil layer of 1 m, with a field capacity value of 0.1 (vol/vol units: 0.1 m^3/m^3) and a net recharge rate of 0.002 m/d. By increasing soil adsorption and air partition, the travel time increases due to the increasing value of RF. As t_c and RF are directly proportional, they can be used interchangeably and have the same limitations.

Another simple index was the *leachability potential* Lp (Laskowski et al., 1982):

$$L_p = S_W \; t_{1/2} / \left(P \; K_{oc}\right) \tag{1.130}$$

where S_W and P are the water solubility and vapor pressure, respectively, (both referred to the same aggregation state) and $t_{1/2}$ the half-life of the chemical in soil. Soil properties and water leaching rate are not included in this index, but chemical reaction is considered. The index is only based on the intrinsic properties of the substance.

The *Attenuation Factor* (Rao et al., 1985) was introduced to combine reactivity with mobility in the retardation factor approach for pesticides. The travel time t_r (d) in Equation 1.127 was used in the following expression, assuming that pesticide degradation follows first-order kinetics, with a rate constant k (1/d):

$$M_2 = M_0 \exp\left(-t_r \; k\right) \tag{1.131}$$

where M_2 is the mass (e.g., moles) of active ingredient able to reach the water table after crossing the root zone and the vadose zone, and M_0 is the mass of active ingredient applied to the surface soil (in the same units as M_2). The attenuation factor AF is defined as:

$$AF = M_2 / M_0 = \exp\left(-t_r \; k\right) \tag{1.132}$$

and indicates the fraction of pesticide able to reach the water table at the time t_r. This attractive index indicates the mass fraction of chemical remaining in the soil when contamination of water begins. It works quite well at relatively high values (10^{-1} to 10^{-2}); however, for chemicals which are not good leachers (a high t_r, sometimes combined with a high k), the AF values show a tendency to produce very high negative exponents (e.g., 10^{-50}, 10^{-90}, etc.).

The MR *(residual mass)* index of the Jury research group (Jury et al., 1987) gives the fraction of chemical reaching a given depth (e.g., 3 m), by an approach similar to that of Equation 1.132, but with the reaction rate varying with soil depth.

More recently, another index, the *groundwater ubiquity score,* GUS, was introduced by Gustafson (1989),

$$GUS = Log\left(t_{1/2}\right)\left[4 - Log\left(K_{oc}\right)\right] \tag{1.133}$$

where $t_{1/2}$ is the soil half-life (d) and K_{oc} is the organic carbon/water partition coefficient (L/kg). The number 4 is an arbitrary value selected to permit separation of pesticides into three categories: leachers (GUS > 2.8), nonleachers (GUS < 1.8), and transition (1.8 < GUS < 2.8). The model can be applied to nonpolar and polar chemicals, after calculating the K_{oc} corresponding to the actual K_D value ($K_{oc} = K_D/f_{oc}$).

The last arrival in the family of leaching index is that of Bacci and Gaggi (1993), derived from the *surface soil model* by Mackay (1991). In Mackay's approach, the chemical is applied to the soil at a given rate (e.g., 1 kg/ha), corresponding to a load M (mol), and it is assumed that the substance is homogeneously distributed throughout a defined soil volume (e.g., 1 m^2 surface soil for a depth of 0.1 m). The equilibrium partitioning of the chemical in all the soil components considered (air, water, organic matter, and mineral matter) is then calculated from the physical and partition properties of the chemical, after defining volumes (m^3) and densities (kg/L) of the compartments by simple fugacity calculations; at equilibrium,

$$M = C_1V_1 + C_2V_2 + C_3V_3 + C_4V_4 \tag{1.134}$$

where M (mol) is the load of the chemical, and C and V the concentration (mol/m^3) and volume (m^3), respectively, of air (subscript 1), water (subscript 2), organic matter (subscript 3), and mineral matter (subscript 4).

According to the fugacity approach, $C_i = f_iZ_i$. In other words, the concentration of the contaminant in compartment i (mol/m^3) is equal to the product of the fugacity f (Pa) and the fugacity capacity Z [mol/(m^3·Pa)]. At equilibrium, the fugacity is the same for all communicating compartments and Equation 1.134 becomes:

$$M = f\left(Z_1V_1 + Z_2V_2 + Z_3V_3 + Z_4V_4\right) \tag{1.135}$$

To calculate f, Z values are obtained by current procedures. For air, $Z_1 = 1/RT$, where R is the gas constant, 8.314 Pa·m^3/(mol K); for water, $Z_2 = C/P$, with C indicating the water solubility (mol/m^3) and P the vapor pressure (Pa) of the chemical; for soil organic matter, $Z_3 = Z_2\,\rho_s\,0.56\,K_{oc}$, where ρ_s is the soil bulk density (kg/L), K_{oc} is the organic carbon/water partition coefficient (L/kg), and 0.56 is a factor introduced on the assumption that the organic matter is 56% organic carbon; for mineral matter, $Z_4 = Z_2 \rho_s K_{pm}$, where K_{pm} is the mineral matter/water partition coefficient (L/kg) required for the application of the model to polar substances. The K_{pm} value for polar chemicals can be obtained by experimental measurement of soil/water partition coefficient (K_D, L/kg) as follows:

$$K_{pm} = K_D / f_{pm} \tag{1.136}$$

where f_{pm} is the volumetric fraction of mineral matter. The fugacity f (Pa) is then obtained from:

$$f = M/(Z_1V_1 + Z_2V_2 + Z_3V_3 + Z_4V_4) \tag{1.137}$$

and the partition of the substance is calculated from the products fZV for each one of the four soil components.

The Z values are then used to calculate transport and reaction parameters, D [mol/(Pa·h)]. For water transport, the water leaching rate (mm/d) is transformed in water flow rate G_L (m^3/h) and combined with the water Z value, Z_2, to obtain the leaching D value, D_L:

$$D_L = G_L Z_2 \tag{1.138}$$

For air transport, according to Jury et al. (1984), three components are involved in the overall volatilization D value, D_V:

- The air boundary layer diffusion parameter, $D_E = Ak_vZ_1$, where A is the soil area (m^2) and k_v the ratio of the diffusivity in air of the substance, B_A, typically 0.0179 m^2/h, to air boundary layer thickness (0.00475 m).
- The air diffusion D value, D_A, obtained as follows: $D_A = B_{EA} A Z_1/Y$ where Y is the diffusion path length (for thin soil layers, e.g., 5 to 10 cm, half the thickness can be selected and B_{EA} is the effective diffusivity obtained from the molecular diffusivity in air of the substance, B_A multiplied by $v_1^{10/3}/(v_1 + v_2)^2$ where v_1 and v_2 are the volume fractions of air and water, respectively, according to the Millington-Quirk model (Shearer et al., 1973).
- The water diffusion D value, D_W, is calculated by a similar approach: $D_W = B_{EW} A Z_2/Y$ where $B_{EW} = B_W v_2^{10/3}/(v_1 + v_2)^2$ and B_W is the molecular diffusivity in water, 4.3×10^{-5} m^2/d, or 0.179×10^{-5} m^2/h.

The total diffusion D value, D_V, is given as follows (Mackay, 1991):

$$1/D_V = 1/D_E + 1/(D_A + D_W) \tag{1.139}$$

Finally, the reaction D value, D_R, is considered; this is obtained as indicated below:

$$D_R = k_R \, V_T Z_T \tag{1.140}$$

where k_R is the overall first-order reaction rate constant, 1/h, V_T the soil volume and Z_T the sum of the Z values for air, water, organic and mineral matter. From this system, the homogeneously distributed chemical will escape in the three ways considered (volatilization, leaching, and reaction) with initial rates (mol/h) as follows:

- Volatilization: fD_V
- Leaching: fD_L
- Reaction: fD_R

The total dissipation or disappearance rate of the chemical from the soil will be *f*DT, with $D_T = D_V + D_R + D_L$; from Equation 15, the single process rate constants (1/h) are:

- Volatilization: $k_V = D_V/(V_T Z_T)$
- Leaching: $k_L = D_L/(V_T Z_T)$
- Reaction: $k_R = D_R/(V_T Z_T)$

The surface soil model is able to provide the relative significance of the different mechanisms of soil chemical depletion, the initial flow in each direction, and the law of clearance of the contaminated soil layer:

$$M_t = M_o \exp\left(-k_{dis} \, t\right) \tag{1.141}$$

where M_t is the mass (mol) of contaminant remaining in the soil layer at time t (h), M_o the initial load (mol), t the elapsed time (h), and $k_{dis} = k_V + k_L + k_R$ the disappearance rate constant (1/h). Mackay suggests that the calculation of the fraction remaining in the soil after a selected time interval could be used to evaluate the potential for groundwater contamination.

A little different approach was suggested by Bacci and Gaggi (1993): a first-order decay is applied to the disappearance of the chemical from the soil (Equation 1.141) and, after a given time interval, the fraction of the chemical leaving the soil by leaching is calculated and used as the leaching index.

The total amount of leaching from the soil layer, LM_t, in mass/surface units (the same used for the load, M_o) in the time interval $t_2 - t_1$ is as follows:

$$LM_t = \int_{t1}^{t2} J_{L(t)} dt \tag{1.142}$$

where $J_{L(t)}$ is the leaching flux [mol/(m^2·h)] at time t (h). According to the assumption of first-order kinetics, $J_{L(t)}$ will show the same decay as M_o, according to Equation 1.141:

$$J_{L(t)} = J_{L(o)} \exp(-k_{dis}\, t) \tag{1.143}$$

where $J_{L(o)} = f\, D_L/A$ is the initial leaching flux [mol/(m^2·h)] from the soil surface A (m^2). Combining Equations 1.142 and 1.143 and calculating the integral, the mass per surface unit LM_t leaching in the time interval $t_2 - t_1$ will be:

$$LM_t = J_{L(o)}\, 1/k_{dis} \left[\exp(-k_{dis} t_1) - \exp(-k_{dis} t_2)\right] \tag{1.144}$$

When referred to the same surface, the LM_t value (mol/m^3) may be expressed as the mass fraction of the initial chemical load, M_0, by the ratio LM_t/M_0. In relation to a standardized soil layer and time interval, this ratio can be used as an index of leaching potential *(leaching index)*.

The application of this approach to the calculation of the leaching index LM_t/M_0 of some pesticides is illustrated in Table 1.10, selecting 1 year (8760 h) as the time interval.

Soil properties:
- Surface: 1 m^2
- Depth: 0.1 m
- Water leaching rate: 2 mm/d $\rightarrow 8.33 \times 10^{-5}$ m/h
- Soil porosity: 0.5 m^3/m^3
- Soil bulk density: 1.3 kg/L
- Field capacity: 0.1 m^3/m^3
- Air filled porosity: 0.4 m^3/m^3
- Organic carbon mass fraction: 0.01 g/g
- Mineral matter volumetric fraction: 0.5 m^3/m^3

Required properties of chemicals:
- Molar mass, g/mol
- Vapor pressure, Pa
- water solubility, mol/m^3
- K_{oc}, L/kg or K_{pm}, L/kg
- Reaction rate constant, k_R, 1/h

Load: 1 kg/ha $\rightarrow$ 0.1 g/m$^2 \rightarrow$ 0.1/molar mass mol/m^2

The index is based on a *constant time* interval. The selection of 1 year for the time interval is based on the consideration that this duration is sufficient to eliminate more than 95% of the residues of most chemicals from the soil layer (both in the simulation and under field conditions). The range of the index is limited to four orders of magnitude (from 1 to 10^{-4}), sufficient to discriminate leachers from transition chemicals and nonleachers. The division was arbitrary, on the basis of substances which are known to leach under field conditions. Thus, substances with values of LM_t/M_0 from 1 to 1×10^{-1} can be regarded as good leachers, while the remaining are unlikely to be found in groundwater.

The ranking model illustrated above is applicable to both nonpolar and polar chemicals via selection of the mineral mass or the organic carbon-based soil/water

Table 1.10. Examples of Calculated Leaching Indices, LM_t/M_o, at 1 year (8760 h)

Known leachers		Transition		Nonleachers	
Chemical	**LM_t/M_o**	**Chemical**	**LM_t/M_o**	**Chemical**	**LM_t/M_o**
Tebuthiuron	9.1×10^{-1}	Carbaryl	7.1×10^{-2}	Benfluralin	9.4×10^{-3}
Carbofuran	8.0×10^{-1}	Triallate	7.0×10^{-2}	Endosulfan	9.4×10^{-3}
Atrazine	5.7×10^{-1}	Desdemipham	4.5×10^{-2}	Triphenyltin-OH	7.4×10^{-3}
Acephate	4.1×10^{-1}	Pendimethalin	3.8×10^{-2}	Diflubenzuron	2.4×10^{-3}
Bentazon	3.9×10^{-1}	Azinphos-methyl	2.3×10^{-2}	Methyl-bromide	1.8×10^{-3}
2,4-D	3.3×10^{-1}	Trifluralin	1.6×10^{-2}	Cypermethrin	7.1×10^{-4}
Alachlor	1.7×10^{-1}	Chlorpyrifos	1.1×10^{-2}	Diquat	2.7×10^{-4}

Note: Properties of the chemicals were taken from the SCS/ARS/CES Pesticide Properties Database (Wauchope et al., 1992). For polar chemicals, K_{pm} was calculated from K_{oc} data normalized to 0.01 organic carbon mass fraction as follows: $K_{oc} \times 0.01 = K_D$ and $K_{pm} = K_D/f_{pm}$ (From Bacci and Gaggi, 1993.)

partition coefficient. The information it provides is the mass fraction leaving the soil layer by leaching in a selected time interval, and it takes mobility and transformation processes occurring in the soil into account.

Soil and rocks are never completely impermeable; wherever groundwater exists, the use of leaching indices in combination with groundwater vulnerability maps (based on soil permeability) may help in the protecting this essential resource, by indicating those chemicals which could not be compatible with the required water quality objective.

1.5. THE ROLE OF FIELD STUDIES

In previous discussions, the focus has been mainly on predictive tools, due to their recent development and relatively restricted application. Field work will remain a milestone, especially when combined with some use of models and, particularly, of evaluative models. Evaluative models are more suitable in producing rankings, rather than data directly transferable to a real situation. However, the indications from sometimes very simple models may greatly aid in planning field work (*where, when, how* and *why* collect samples). A wrong hypothesis may be revealed by field data, and modified until an agreement is reached. Field studies are also needed to validate or to calibrate model predictions in order to point out key processes determining the environmental fate of chemicals. Once these are known, by means of an integrated approach based on physicomathematical models, laboratory models and field research, temporal and spatial sampling grids may be prepared and applied for routine and legislative controls.

A recommendation: when studying contaminants in the environment, never forget to have in mind an idea of the load and pathways linking the sources with the system under examination. The reason for this statement originates from the fact that modern technology provides continuously new analytical instruments, always more perfect, rapid, accurate, computerized, robotized, relatively cheap, able to produce a lot of data, even with very little human help. Unfortunately, environmental chemistry needs a human contribution not only for correctly carrying out the analytical work (essential but not sufficient), but also for planning the investigation and, then for the interpretation of results.

A little story to conclude this first section. Once, a friend of mine, stressed by a tremendous (and unjustified) pollution fear, asked me to check the water quality of his little lake. The lake receives runoff waters from a narrow agricultural area, where some pesticides are currently applied. He brought to the laboratory a bottle of water, requiring the analysis of "all possible contaminants". I indicated two alternatives: the first one was starting from the analysis of the problem, leaving the analysis of water to a further moment, after studying loads and properties of the chemicals; the second alternative was to carry out a long series of analyses, looking for all possible contaminants, but with the serious risk to completely dry his lake (and his money), before accomplishing the study!

REFERENCES

Alzieu, C., TBT detrimental effects on oyster culture in France, in *Oceans '86 Proceedings*, Vol. 4, International Organotin Symposium. The Institute of Electrical and Electronics Engineers, New York, pp. 1130–1134 (1986).

Alzieu, C., Michel, P., Tolosa, I., Bacci, E., Mee, L.D., and Readman, J.W., Organotin compounds in the Mediterranean: a continuing cause for concern, *Mar. Environ. Res.*, 32: 261-270 (1991).

ASTM, *Standard Practice for Evaluating Environmental Fate Models of Chemicals. 1978-84.* American Society for Testing and Materials, Philadelphia, PA (1984).

Bacci, E., Mercury in the Mediterranean, *Mar. Pollut. Bull.,* 20: 59-63 (1989).

Bacci, E., Baldi, F., Bargagli, R., and Gaggi, C., Recovery trends in a mercury-polluted marine area, in *Proceedings* of the *FAO/UNEP/WHO/IOC/IAEA Meeting on the Biogeochemical Cycle of Mercury in the Mediterranean,* Siena, August 27-31, 1984, FAO Fisheries Report No. 325 Supplement, Rome, Italy, pp. 20-28, (1984).

Bacci, E., Calamari, D., Gaggi, C., and Vighi, M., An approach for the prediction of environmental distribution and fate of cypermethrin, *Chemosphere*, 16: 1373-1380 (1987).

Bacci, E., Calamari, D., Gaggi, C., and Vighi, M., Bioconcentration of organic chemical vapors in plant leaves: experimental measurements and correlation, *Environ. Sci. Technol.,* 24: 885-889 (1990a).

Bacci, E., Cerejeira, M.J., Gaggi, C., Chemello, G., Calamari, D., and Vighi, M., Bioconcentration of organic chemical vapours in plant leaves: the azalea model, *Chemosphere*, 21: 525-535 (1990b).

Bacci, E., Cerejeira, M.J., Gaggi, C., Chemello, G., Calamari, D., and Vighi, M., Chlorinated dioxins: volatilization from soils and bioconcentration in plant leaves, *Bull. Environ. Contam. Toxicol., 48:* 401-408 (1992).

Bacci, E., and Gaggi, C., Polychlorinated biphenyls in plant foliage: translocation or volatilization from contaminated soils? *Bull. Environ. Contam. Toxicol.,* 35: 673-681 (1985).

Bacci, E., and Gaggi, C., Chlorinated pesticides and plant foliage: translocation experiments, *Bull. Environ. Contam. Toxicol.,* 37: 850-857 (1986).

Bacci, E., and Gaggi, C., Organotin compounds in harbour and marina waters from the Northern Tyrrhenian Sea, *Mar. Pollut. Bull.*, 20: 290-292 (1989).

Bacci, E., and Gaggi, C., Tributyltin in the Leghorn harbour area (Tuscany, Italy): a tentative hazard assessment to aquatic life, in *Proceedings 3rd International Organotin Symposium*, Monaco, April 17-20, 1990, pp. 140-146 (1990).

Bacci, E., and Gaggi, C., Ranking pesticide mobility from surface soil to groundwater: a new leaching index, submitted (1993).

Bacci, E., Renzoni, A., Gaggi, C., Calamari, D., Franchi, A., Vighi, M., and Severi, A., Models, field studies, laboratory experiments: an integrated approach to evaluate the environmental fate of atrazine, *Agr. Ecosyst. Environ.*, 27: 513-522 (1989).

Baughman, G.L., and Lassiter, R.R., Prediction of environmental pollutant concentration, in *Estimating the Hazard of Chemical Substances to Aquatic Life,* Cairns, J., Jr., Dickson, K.L., and Maki, A.W., Eds., American Society for Testing and Materials, Philadelphia, PA, Tech. Publ. 657, pp. 35-54 (1978).

Bidleman, T.F., personal communication (1991).

Bidleman, T.F., Billings, W.N., and Foreman, W.T., Vapor-particle partitioning of semivolatile organic compounds: estimates from field collections, *Environ. Sci. Technol.,* 20: 1038-1043 (1986).

Blank, L.W., Crane, A.J., and Skeffington, R.A., The long-term ecological effects of pollutants: some issues, *Sci. Total Environ.,* 116: 145-158 (1992).

Blum, D.J.W., and Speece, R.E., Determining chemical toxicity to aquatic species, *Environ. Sci. Technol.,* 24: 284-293 (1990).

Blunden, S.J., Hobbs, L.A., and Smith, P.J., The environmental chemistry of organotin compounds, in *Environmental Chemistry,* Bowen, H.J.M., Ed., The Royal Society of Chemistry, London, U.K., pp. 49-77 (1984).

Bodek, I., Lyman, W.J., Reehl, W.F., and Rosenblatt, D.H., *Environmental Inorganic Chemistry. Properties, Processes, and Estimation Methods.* Pergamon Press, New York, (1988).

Braun, H., Metzger, M., and Vogg, H., Neue Erkenntnisse über metallische Schadstoffe in der Luft, *Fresenius Z. Anal. Chem.,* 317: 304-308 (1984).

Calamari, D., Bacci, E., Focardi, S., Gaggi, C., Morosini, M., and Vighi, M., Role of plant biomass in the global environmental partitioning of chlorinated hydrocarbons, *Environ. Sci. Technol.,* 25: 1489-1495 (1991).

Carson, R.L., *Silent Spring.* Houghton Mifflin, Boston (1962).

Cairns, J., Jr., Estimating hazard, *Bioscience,* 30: 101-107 (1980).

Cessna, A.J., and Muir, D.C.G., Photochemical transformations, in *Environmental Chemistry of Herbicides.* Vol. II. Grover, R., and Cessna, A.J., Eds., CRC Press, Boca Raton, FL, pp. 199-263 (1991).

Connell, D.W., *Bioaccumulation of Xenobiotic Compounds.* CRC Press, Boca Raton, FL, (1990).

Connell, D.W., and Hawker, D.W., Predicting the distribution of persistent organic chemicals in the environment, *Chem. Aust.,* 53: 428 (1986).

Davidson, J.M., Reick, C.E., and Santelman, P.W., Influence of water flux and porous materials on the movement of selected herbicides, *Soil. Sci. Soc. Am. Proc.,* 32: 629-633 (1968).

DePablo, R.S., Vapor pressure of dimethyl tetrachloroterephtalate, *J. Chem. Eng. Data,* 26: 237-239 (1981).

Dziegielewski, B., and Baumann, D., The benefits of managing urban water demands. *Environment,* 34: 7-11 and 35-41 (1992).

Eitzer, B.D., and Hites, R., Dioxins and furans in the ambient atmosphere: a baseline study, *Chemosphere*, 18: 593-598 (1989).

Gabrielides, G.P., Alzieu, C., Readman, J.W., Bacci, E., Aboul Dahab, O., and Salihoglu, I., MED POL survey of organotins in the Mediterranean, *Mar. Pollut. Bull.*, 21: 233-237 (1990).

Gaggi C., Chemello, G., and Bacci, E., Mercury vapour accumulation in azalea leaves, *Chemosphere*, 22: 869-872 (1991).

Goldberg, E.D., *The Health of the Oceans*. The Unesco Press, Paris, France, 1976.

Gustafson, D.I., Groundwater ubiquity score: a simple method for assessing pesticide leachability, *Environ. Toxicol. Chem.*, 8: 339-357 (1989).

Harris, J.C., and Hayes, M.J., Acid dissociation constant, in *Handbook of Chemical Property Estimation Methods*, Lyman, W.J., Reehl, W.F., and Rosenblatt, D.H., Eds. American Chemical Society, Washington D.C., pp. (6) 1-28, (1990).

Hartley, G.S., Evaporation of pesticides, *Adv. Chem. Ser.*, 86: 115-134 (1969).

Hartley, G.S., and Graham-Bryce, I.J., *Physical principles of pesticide behaviour*. Vol. 1. Academic Press, London, (1980).

Hawker, D.W., and Connell, D.W., Bioconcentration of lipophilic compounds by some aquatic organisms, *Ecotoxicol. Environ. Saf.*, 11: 184-189 (1986).

Hinckley, D.A., Bidleman, T.F., and Foreman, W., Determination of vapor pressure for nonpolar and semipolar organic compounds from gas chromatographic retention data, *J. Chem. Eng. Data*, 35: 232-237 (1990).

Hope, B.K., Ecological considerations in the practice of ecotoxicology, *Environ. Toxicol. Chem.*, 12: 205 (1993).

Hunt, E.G., and Bischoff, A.I., Inimical effects on wildlife of periodic DDD application to Clear Lake, *Calif. Fish. Game*, 46: 91-106 (1960).

Hutzinger, O., Tulp, M.Th.M., and Zitko, V., Chemicals with pollution potential, in *Aquatic Pollutants: Transformation and Biological Effects*, Hutzinger, O., van Lelyveld, I.H., and Zoeteman, C.J., Eds. Pergamon Press, Oxford, U.K., pp. 13-31 (1978).

Irving, L., Scholander, P.F., and Grinnell, S.W., The respiration of the porpoise, *Tursiops truncatus*, *J. Cell. Comp. Physiol.*, 17: 145-168 (1941).

Jensen, S., The PCB story, *Ambio*, 1: 123-131 (1972).

Jones, D.R., Brill, R.W., Butler, P.J., Bushnell, P.G., and Heieis, M.R.A., Measurement of ventilation volume in swimming tunas, *J. Exp. Biol.*, 149: 491-498 (1990).

Jørgensen, S.E., (Ed.,) *Modelling in Ecotoxicology*. Elsevier, Amsterdam, pp. 15-35 (1990).

Jury W.A., Farmer, W.J., and Spencer, W.F., Behavior assessment model for trace organics in soil: II. Chemical classification and parameter sensitivity, *J. Environ. Qual.*, 13: 567-572 (1984).

Jury, W.A., Focht, D.D, and Farmer, W.J., Evaluation of pesticide groundwater pollution potential from standard indices of soil-chemical adsorption and biodegradation, *J. Environ. Qual.*, 16: 422-428 (1987).

Jury, W.A., Spencer, W.F., and Farmer, W.J., Behavior assessment model for trace organics in soil: I. Model description, *J. Environ. Qual.*, 12: 558-564 (1983).

Karickhoff, S.W., Semiempirical estimation of sorption of hydrophobic pollutants on natural sediments and soils, *Chemosphere*, 10: 833-849 (1981).

Klein, A.W., Goedicke, J., Klein, W., Herrchen, M., and Kördel, W., Environmental assessment of pesticides under Directive 91/414/EEC, *Chemosphere*, 26: 979-1001 (1993).

Koeman, J.H., Ecotoxicology and environmental quality, *Environ. Monitoring Assess.*, 3: 227-228 (1983).

Kurtz, D.A., Ed., *Long Range Transport of Pesticides,* Lewis, Chelsea, MI, (1990).

Laskowski, D.A., Goring, C.A.I., McCall, P.J., and Swann, R.L., Terrestrial environment, in *Environmental Risk Analysis for Chemicals,* Conway, R.A., Ed., Van Nostrand Reinhold, New York, pp. 198-240 (1982).

Laughlin, R.B., Jr., Guard, H.E., and Coleman, W.M., III, Tributyltin in seawater: speciation and octanol-water partition coefficient, *Environ. Sci. Technol.,* 20: 201-204 (1986).

Leifer, A., *The Kinetics of Environmental Aquatic Photochemistry. Theory and Practice,* American Chemical Society, Washington, D.C., (1988).

Lee R.F., Valkirs, A.O., and Selingman, P.F., Fate of tributyltin in estuarine waters, in *Oceans '87 Proceedings,* Vol. 4, International Organotin Symposium. The Institute of Electrical and Electronics Engineers, New York, pp. 1411-1415 (1987)

Leo, A., and Hansch, C., Linear free-energy relationships between partitioning solvent systems, *J. Org. Chem.,* 36: 1539-1544 (1971).

Lyman, W.J., Reehl, W.F., and Rosenblatt, D.H., Eds., *Handbook of Chemical Property Estimation Methods,* American Chemical Society, Washington, D.C., (1990).

Mackay, D., Finding fugacity feasible, *Environ. Sci. Technol.,* 13: 1218-1223 (1979).

Mackay, D., Correlation of bioconcentration factors, *Environ. Sci. Technol.,* 16: 274-278 (1982).

Mackay, D., *Multimedia Environmental Models. The Fugacity Approach,* Lewis, Chelsea, MI, (1991).

Mackay, D., and Paterson, S., Calculating fugacity, *Environ. Sci. Technol.,* 15: 1006-1014 (1981).

Mackay, D., Paterson, S., and Schroeder, W.H., Model describing the rates of transfer processes of organic chemicals between atmosphere and water, *Environ. Sci. Technol.,* 20: 810-816 (1986).

Mackay, D., and Stiver, W., Predictability and environmental chemistry, in *Environmental Chemistry of Herbicides.* Vol. II. Grover, R., and Cessna, A.J., Eds. CRC Press, Boca Raton, FL, pp. 281-297 (1991).

Majewski, M., Desjardins, R., Rochette, P., Pattey, E., Seiber, J., and Glotfelty, D., Field comparison of an eddy accumulation and an aerodynamic-gradient system for measuring pesticide volatilization fluxes, *Environ. Sci. Technol.,* 27: 121-128 (1993).

Majewski, M.S., Glotfelty, D.E., Pau U, K.T., and Seiber, J.N., A field comparison of several methods for measuring pesticide evaporation rates from soil, J. *Environ. Qual.* 24: 1490-1497 (1990).

Major, M.A., Rosenblatt, D.H., and Bostian, K.A., The octanol/water partition coefficient of methylmercuric chloride and methylmercuric hydroxide in pure water and salt solutions, *Environ. Toxicol. Chem.,* 10: 5-8 (1991).

McCrady, J.K., and Maggard, S.P., Uptake and photodegradation of 2,3,7,8-tetrachlorodibenzo-*p*-dioxin sorbed to grass foliage, *Environ. Sci. Technol.,* 27: 343-350 (1993).

McCrady, J., McFarlane, C., and Lindstrom, F.T., The transport and affinity of substituted benzenes in soybean stems, *J. Exp. Bot.,* 38: 1875-1890 (1987).

McDonnell, R., Assessing the impacts: the U.K. situation, *Ambio,* 21: 188-191 (1992).

Metcalf, R.L., Sangha, G.K., and Kapoor, I.P., Model ecosystem for the evaluation of pesticide degradability and ecological magnification, *Environ. Sci. Technol.,* 5: 709-713 (1971).

Mill, T., Data needed to predict the environmental fate of organic chemicals, in *Dynamics, Exposure and Hazard Assessment of Toxic Chemicals,* Haque, R., Ed. Ann Arbor Science, Ann Arbor, MI, pp. 297-322 (1980).

Moriarty, F., *Ecotoxicology. The Study of Pollutants in Ecosystems.* Academic Press, London, U.K., (1983).

Mortimer, C.T., *Reaction Heats and Bond Strengths,* Pergamon Press, New York, (1962).

Munn, R.E., *Environmental Impact Assessment.* SCOPE 5, 2nd ed., John Wiley & Sons, Chichester, U.K., (1975).

Neely, W.B., *Chemicals in the Environment. Distribution, Transport, Fate, Analysis,* Marcel Dekker, New York, (1980).

Neely, W.B., Branson, D.R., and Blau, G.E., The use of the partition coefficient to measure the bioconcentration potential of organic chemicals in fish, *Environ. Sci. Technol.,* 8: 1113-1115 (1974).

Newman, M.C., Regression analysis of log-transformed data: Statistical bias and its correction, *Environ. Toxicol. Chem.,* 12: 1129-1133 (1993).

Paterson, S., Mackay, D., Bacci, E., and Calamari, D., Correlation of the equilibrium and kinetics of leaf/air exchange of hydrophobic organic chemicals, *Environ. Sci. Technol.,* 25: 866-871 (1991).

Phillips, D.J.H., and Segar, D. A., Use of bio-indicators in monitoring conservative contaminants: programme design imperatives, *Mar. Pollut. Bull.,* 17: 10-17 (1986).

Rao, P.S.C., Hornsby, A.G., and Jessup, R.E., Indices for ranking the potential for pesticide contamination of groundwater, *Proc. Soil Crop Sci. Soc. Fla.,* 44: 1-8 (1985).

Rechsteiner, C.E., Jr., Heat of vaporization, in *Handbook of Chemical Property Estimation Methods,* Lyman, W.J., Reehl, W.F., and Rosenblatt, D.H., Eds. American Chemical Society, Washington, D.C., pp. 13.1-13.28 (1990).

Renzoni, A., Bacci, E., and Falciai, L., Mercury concentration in the water, sediment and fauna of an area of the Tyrrhenian coast, *Rev. Intern. Océanogr. Méd.,* 31-32: 17-45 (1973).

Renzoni, A., Chemello, G., Gaggi, C., Bargagli, R., and Bacci, E., Methylmercury in deep-sea organisms from the Mediterranean, in *Proceedings of the FAO/UNEP/IAEA Consultation Meeting on the Accumulation and Transformation of Chemical Contaminants by Biotic and Abiotic Processes in the Marine Environment,* Gabrielides, G.P., Ed. La Spezia (Italy) September 24-28, 1990. UNEP, MAP Technical Reports Series No. 59, Athens, Greece, pp. 303-318 (1991).

Ridgway, S.H., Scronce, B.L., and Kanwisher, J., Respiration and deep diving in a bottlenose porpoise, *Science,* 166: 1651-1654 (1969).

Ross, L.J., Nicosia, S., McChesney, M.M., Hefner, K.L., Gonzalez, D.A., and Seiber, J.N., Volatilization, off-site deposition, and dissipation of DCPA in the field, *J. Environ. Qual.,* 19: 715-722 (1990).

Royal Commission, *Tackling Pollution Experience and Prospects.* Royal Commission on Environmental Pollution, Tenth Report. Her Majesty's Stationery Office, London, U.K., (1984).

Ryan, J.A., Bell, R.M., Davidson, J.M., and O'Connor, G.A., Plant uptake of non-ionic organic chemicals from soils, *Chemosphere,* 17: 2299-2323 (1988).

Schwarzenbach, R.P., Gschwend, P.M., and Imboden, D.M., *Environmental Organic Chemistry,* John Wiley & Sons, New York, (1993).

Schroll, R., and Scheunert, I., A laboratory system to determine separately the uptake of organic chemicals from soil by plant roots and by leaves after vaporization, *Chemosphere,* 24: 97-108 (1992).

Schwarzenbach, R.P., Stierli, R., Lanz, K., and Zeyer, J., Quinone and iron porphyrin mediated reduction of nitroaromatic components in homogeneous aqueous solution, *Environ. Sci. Technol.,* 24: 1566-1574 (1990).

Scow, K.M., Rate of biodegradation, in *Handbook of Chemical Property Estimation Methods,* Lyman, W.J., Reehl, W.F., and Rosenblatt, D.H., Eds. American Chemical Society, Washington, D.C., pp. 9.-1 to 9.-85 (1990).

Shearer, R.C., Letey, J., Farmer, W.J., and Klute, A., Lindane diffusion in soil, *Soil Sci. Soc. Am. Proc.,* 37: 189-193 (1973).

Shoichi, O. and Sokichi, S., The 1-octanol/water partition coefficient of mercury, *Bull. Chem. Soc. Jpn.,* 58: 3401-3402 (1985).

Spencer, W.F., and Cliath, M.M., Pesticide volatilization as related to water loss from soil, *J. Environ. Qual.,* 2: 284-289 (1973).

Stang, P.M., and Selingman, P.F., Distribution and fate of butyltin compounds in the sediment of San Diego bay, in *Oceans '86 Proceedings,* Vol. 4, International Organotin Symposium. The Institute of Electrical and Electronics Engineers, New York, pp. 1256-1261 (1986).

Stiver, W., and Mackay, D., Predictability and environmental chemistry, in *Environmental Chemistry of Herbicides.* Vol. II, Grover, R., and Cessna, A.J., Eds. CRC Press, Boca Raton, FL, pp. 281-297 (1991).

Stumm, W., Schwarzenbach, R.P., and Sigg, L., From environmental analytical chemistry to ecotoxicology—A plea for more concepts and less monitoring and testing, *Angew. Chem.,* 22: 380-389 (1983).

Suntio, L.R., Shiu, W.Y., Mackay, D., Seiber, J.N., and Glotfelty, D., Critical review of Henry's law constants for pesticides, *Rev. Environ. Contam. Toxicol.,* 103: 1-59 (1988).

Travis, C.C., and Hattemer-Frey, H.A., Uptake of organics by aerial plant parts: a call for research, *Chemosphere,* 17: 277-283 (1988).

Trevisan, M., Capri, E., and Ghebbioni, C., L'uso di indici per prevedere la contaminazione delle acque, *Acqua & Aria,* 9: 863-874 (1991).

Trouton, F., IV. On molecular latent heat, *Phil. Mag.,* 18: 54-57 (1884).

Truhaut, R., Ecotoxicology—A new branch of toxicology: a general survey of its aims, methods, and prospects, in *Ecological Toxicology Research,* McIntyre, A.D., and Mills, C.F., Eds. Plenum Press, New York, pp. 3-23 (1975).

Tyler Miller, G., Jr., *Resource Conservation and Management.* Wadsworth Publishing, Belmont, CA, (1989).

van Esch, G.J., Aquatic pollutants and their potential biological effects, in *Aquatic Pollutants: Transformation and Biological Effects,* Hutzinger, O., van Lelyveld, I.H., and Zoeteman, C.J., Eds. Pergamon Press, Oxford, U.K., pp. 1-12 (1978).

Veith, G.D., DeFoe, D.L., and Bergstedt, B.V., Measuring and estimating the bioconcentration factor of chemicals in fish, *J. Fish. Res. Board Can.,* 36: 1040-1048 (1979).

Wagenet, R.J., and Rao, P.S.C., Modeling pesticide fate in soils, in *Pesticides in the Soil Environment: Processes, Impacts, and Modeling,* Cheng, H.H., Ed. Soil Science Society of America, Inc., Madison, WI, pp. 351-399 (1990).

Wauchope, R.D., Buttler, T.M., Hornsby, A.G., Augustijn-Beckers, P.W.M., and Burt, J.P., The SCS/ARS/CES pesticide properties database for environmental decision-making, *Rev. Environ. Contam. Toxicol.,* 123: 1-155 (1992).

Weber, J.B., Mechanisms of adsorption of *s*-triazines by clay colloids and factors affecting plant availability, *Residue Rev.,* 32: 93-130 (1970).

Welhouse, G.J., and Bleam, W.F., Atrazine hydrogen-bonding potentials, *Environ. Sci. Technol.,* 27: 494-500 (1993).

Wells, P.G., and Côté, R.P., Protecting marine environmental quality from land-based pollutants, *Marine Policy,* Jan. 1988, pp 9-21 (1988).

Worthing, C.R., and Hance, R.J., *The Pesticide Manual,* 9th ed. The British Crop Protection Council, Farnham, U.K., (1991).

Wright, D.A., Less *"Where?",* more *"How?"* and *"Why?", Mar. Pollut. Bull.,* 16: 432-435 (1985).

Yalkowsky, S.H., Estimation of entropies of fusion of organic compounds, *Ind. Eng. Chem. Fund.,* 18: 108-111 (1979).

Yalkowsky, S.H., Valvani, S.C., and Mackay, D., Estimation of the aqueous solubility of some aromatic compounds, *Residue Rev.,* 85: 43-55 (1983).

Yin, C., and Hassett, J.P., Gas partitioning approach for laboratory and field studies of mirex fugacity in water, *Environ. Sci. Technol.,* 20: 1313-1217 (1986).

Zepp, R., and Baughman, G., Prediction of photochemical transformation of pollutants in the aquatic environment, in *Aquatic Pollutants. Transformation and Biological Effects,* Hutzinger, O., van Lelyveld, I.H., and Zoeteman, B.C.J., Eds. Pergamon Press, Oxford, U.K., pp. 237-263 (1978).

And this our life, exempt from public haunt, finds tongues in trees, books in running brooks, sermons in stones, and good in everything.

William Shakespeare

SECTION 2 Environmental Toxicology

2.1. INTRODUCTION

The first section dealt with approaches to evaluate the environmental fate of organic contaminants. Living organisms are included among the environmental compartments; they may enter in contact with and take up contaminants. The contact refers to the *exposure* of the biological system; exposure is quantified by the concentration in the medium (e.g., for a fish, the concentration in water), together with an indication of the time of exposure duration. The intake refers to the *dose* which is currently quantified by the ratio of the mass of chemical divided by the product of a unitary mass of organism and time; for example, mg chemical/(kg body weight × day), or for *single dose*, more simply, mg chemical/kg body weight.

A definition of the terms *exposure* and *dose* for humans was adopted by UNEP/WHO (1977):

> "The exposure to a given pollutant is a measure of the contact between the pollutant and the outer or inner (e.g., alveolar surface or gut) surface of the human body. Is is usually expressed in terms of concentrations of the pollutant in the medium (e.g., ambient air and food) interfacing with body surfaces. Once absorbed through body surfaces, the pollutant gives rise to doses in various organs and tissues. Doses are measured in terms of concentrations in the tissues. Records of exposure and dose should include an indication of the time and frequency at which an individual is subject to them."

This definition may apply to other living organisms.

It is assumed that exposure and dose are directly proportional. Consequently, dose-effect relationships may be transformed into the correlation exposure–effect by means of a proportionality constant.

As stated by Paracelsus (Deichmann et al., 1986), *"Was ist das nit gifft ist: alle ding sind gifft/und nichts ohn gifft/Allein die dosis macht das ein ding kein gifft ist"*, or "all substances are poisons; there is not one which is not a poison; the right dose makes the difference between a poison and a remedy" (Doull, 1987). With this statement, or the *dose-effect concept*, toxicology began. It may be observed that the original work by Paracelsus, written in german-swiss in the first half of the 16th century, was translated into Latin a few decades later in the famous sentence: *"Dosis sola facit venenum"*, meaning that substances are not *per se* poisons, but the dose is able to transform them into poisons. This misinterpretation of Paracelsus' original idea follows the views of the physicians of that period who were more convinced of the beneficial effects of their drugs than of possible adverse effects. In the case of environmental contamination, the original approach of Paracelsus is probably more appropriate, particularly in the case of biocidal chemicals spread in the environment for the protection of materials (e.g., antifouling treatments, wood preservation), crops (e.g., pesticide application), and humans and animals (e.g., vector control operations). These substances are produced and applied because of their biocidal activity, or toxicity, or the ability to produce damage on some more or less restricted group of organisms, so they are poisons and potentially toxic. Only the dose can make them safe. Environmental toxicology aims to measure the toxicity of substances and to indicate the levels of exposure and doses at which these can be considered as safe or, better, at which the risk of environmental damage is kept below a selected level.

In Section 1, the following definition was proposed for environmental pollution:

> *Environmental pollution occurs when contamination produces measurable damage to living organisms, populations, or biological communities.*

Environmental toxicology concerns the identification and quantification of possible adverse effects on living organisms due to the exposure to environmental contaminants. The roots of environmental toxicology are in the classic toxicology: there are no substantial differences, apart from the nature of the toxic agent, in the identification and quantification of the *toxic action* where analogous approches are applied to understand the biochemical, physiological and biological mechanisms influenced by the *exposure* to an *environmental contaminant*, or to a *mixture of contaminants*. Also in the quantification, there are many similarities: current methods from classic toxicology are applied for the *evaluation of acute and chronic toxicity* in environmental toxicology.

The reasons which support the differentiation of an environmental branch of toxicology are:

- The need to direct the study of toxicity to species other than man (all living organisms may be endangered by pollutants)
- The need to evaluate toxicity at the level of biological community over an appropriate time scale

An additional complication, characterizing environmental toxicology, is the evaluation of exposure: this implies knowledge on the environmental distribution and fate of contaminants, which is the field of *environmental chemistry*.

The combination of environmental chemistry and environmental toxicology may be called *ecotoxicology* (Truhaut, 1975; see also Miller, 1978) which final aim is, still now, a challenge and consists in the production of scientific criteria for *ecological risk assessment*. In this view, the role of environmental toxicology can be restricted to the investigation of *"safe" exposure levels* or to the evaluation of levels corresponding to an *acceptable risk for biological systems at different levels of organization, including biological communities*.

2.2. CLASSIC TOXICOLOGY

As stated by John Doull (1984), *"Toxicology, like medicine, is both a science and an art."* This is particularly true when data observed in a group of selected animals in a given exposure interval and controlled conditions (*science*) is extrapolated to predict effects at lower exposures, or inferred to a larger group of similar organisms (i.e., the species) or to other species, even on the basis of larger temporal scales (*art?*). The concern of toxicology is how to produce scientifically sound data, while the more *artistic* topic concerning hazard and risk assessment will be discussed in Section 3, where toxicological knowledge is extrapolated, inferred, and combined with exposure or dose data.

Toxicology, after the first key concept of *dose-effect* relationships, received a second fundamental impulse during the past century when the practice of measuring the *incidence of adverse effects* (i.e., the *biological response*) in groups of test animals of one species was introduced by Orfila to predict responses in other species. In the middle of the 19th century, Orfila founded modern toxicology (Doull, 1987), or the science of poisons, combining chemistry with jurisprudence (this part of toxicology is now called *forensic toxicology*).

2.2.1. Effects and Responses

The principle of dose-effect was introduced by Paracelsus; Orfila suggested the way to measure effects by means of the biological response and, later, Trevan (1927) developed the *sigmoid curve* and suggested the median lethal dose (LD_{50}) approach. Later, Bliss (1935) introduced the PROBIT method to calculate LD_{50}.

Effect and response, in toxicology, have two different meanings (WHO, 1978):

- Effect indicates the damage, or the biological function compromised by the action of the toxic (e.g., survival, motility, growth rate)
- Response is the portion of exposed organisms showing a determinate effect due to the toxic action (% incidence)

Effects and responses depend on several factors, other than those related to the tested biological species and treated group characteristics. These may be related to the test species or to some abiotic factor; in the first case, the factors that modify toxicity include health conditions, eventual acclimatization, genetic variability, age, sex (if applicable), and, in humans, lifestyle. Abiotic factors relate to temperature, oxygen availability, and type of diet; in aquatic toxicological tests: pH, hardness and salinity of water, and concentration of particulate matter. Detailed discussions on these factors can be found in available literature (Sprague, 1985; Carlson, 1987; Cox et al., 1992).

Assuming that the test species were correctly selected and the experiment conducted in an optimal way, the main factors affecting effects and responses are as follows:

- The *exposure level*, which can be espressed by the *dose*, or the quantity administered to each test organism per unit weight of the organisms (i.e., mg/kg body weight), as the toxicant was homogeneously diluted in the body of the receiving organisms; an alternative to the dose is the *concentration* of the toxicant in the medium surrounding the test species (e.g., air in phytotoxicity studies with vapors, or water in aquatic toxicology tests).
- The *exposure time,* or *duration*, indicating the time of treatment of test organisms.

2.2.2. Measurement of Effects

The exposure to the toxic agent for a living organism occurs when there is a "collision" between the two. According to the structure and physiology of the organism, this may happen in different ways: contact with the organism membranes and penetration inside the cells. In mammals, this may occur by inhalation, ingestion, dermal- and mucous-tissue contact. There are also other possibilities of exposure, particularly for drugs; for example, by direct injection (ipodermal, intravenous, or intraperitoneal).

For a given effect (e.g., lethality), the responses (expressed in percent incidence) are quantitatively related to both the exposure level and the exposure time. In Figure 2.1., a schematic representation of this correlation is given. The correlation illustrated in Figure 2.1 is based on three assumptions:

- The substance used in the toxicological test was causing the observed effect
- The biological effects and relative responses vary as a function of the dose, or of the exposure level (i.e., concentration in the medium)
- The dose, or the quantity taken up by the test organisms, is directly proportional to the exposure level (e.g., the concentration of toxicant in the surrounding water or air)

As illustrated in Figure 2.1., from the three-dimensional response-concentration-duration surface, three different types of correlations may be obtained, keeping the third variable constant. These include the fundamental tools for measuring toxicity.

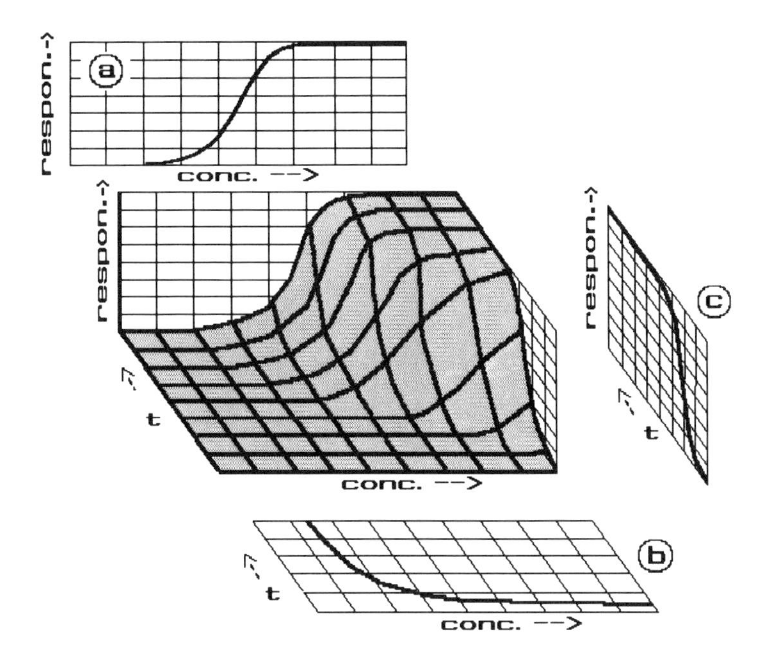

Figure 2.1. The *response–concentration–duration surface.* Sections a, b, and c indicate the following toxicity curves, in the order: the classic response–concentration, time–concentration, and response–time. (Modified from Hartung, 1987. With permission.)

2.2.2.1. Acute Toxicity

Acute toxicity concerns effects occurring after a short exposure time: 24 to 96 h; shorter exposures (e.g., 30 min) may be applied in particular tests, such as those for the resistance to strong acid vapors (e.g., hydrogen fluoride and hydrogen chloride; Stavert et al., 1991) or to simulate industrial accidents. The typical endpoint is lethality. Figure 2.1a indicates the classic *concentration* (or *dose*)-*response curve* obtained when treating groups of organisms with different concentrations or different doses of the same chemical and observing the effect over a fixed period of time (typically 24 h). The curve could better be called response-dose, as the response depends on the dose (independent variable) and not vice versa. The response-dose curve appears to be sigmoid, with a more or less wide linear portion in the middle, and is not generally symmetrical, especially when concentrations (or doses) are in arithmetic progression, due to a slower increase in responses with increasing exposure level when approaching the maximum (100% incidence). The reason for the sigmoid trend, as in cumulative frequency distribution functions, relies on the fact that the observed responses at each dose (or concentration level) actually are *cumulative.*

Of particular concern, in short-term experiments, is the 50% incidence and the concentration or dose provoking this response. If the effect is the death of treated organisms, this concentration is the *median lethal concentration*, LC_{50}, or the

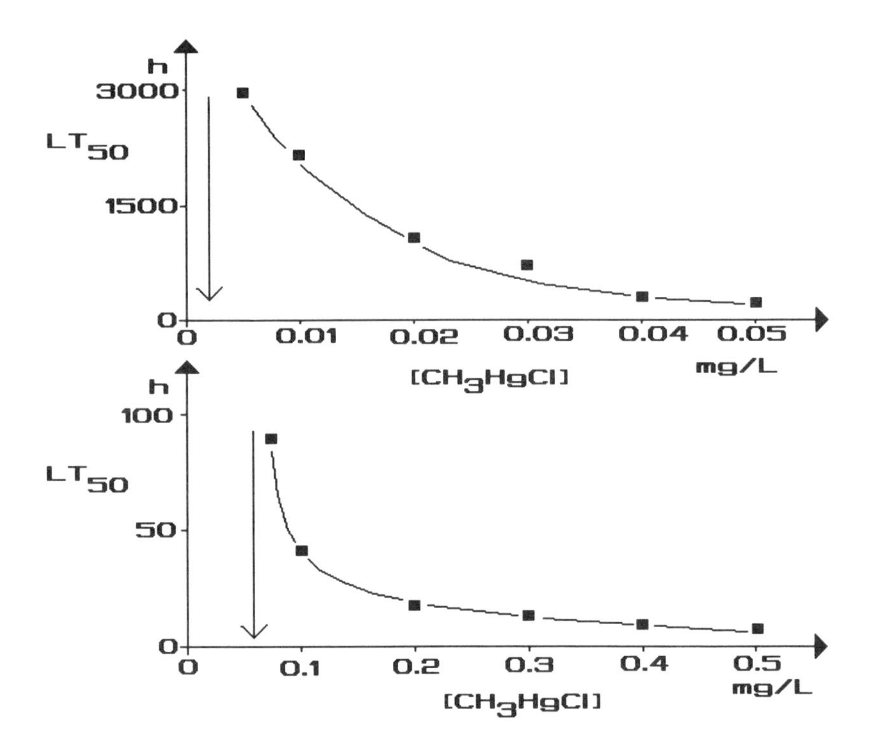

Figure 2.2. Median lethal times and incipient lethal threshold concentrations (indicated by the arrows) in *Carassius auratus* treated with methylmercury. (Unpublished results.)

median lethal dose, LD_{50}. If the effect is the inhibition of some important biological function, the concentration corresponding to a 50% response is called IC_{50} (*median inhibition concentration*). Concentrations causing different effects on 50% of treated organisms are called *median effective concentrations*, EC_{50} (Rand and Petrocelli, 1985).

Other important reference points, particularly in chronic toxicity studies (where only the initial part of the sigmoid curve is studied), are those indicating lower response levels, such as the EC_{01}, which is the effective concentration for 1% of treated organisms.

Another approach consists of measuring the time needed for the occurrence of a particular response; for example, the death of 50% of treated organisms. This can be obtained by keeping the incidence of a selected effect constant and experimentally observing the duration to reach this response as a function of the concentration or dose (Figure 2.1b). In this case, the duration corresponds to the survival time for 50% (*median survival time*) of treated organisms, and to the lethal time of the other 50%; this duration is often called *median lethal time*, LT_{50} (Figure 2.2). The third possibility (Figure 2.1c) is when, with a constant concentration or constant dose, the variation of the response with exposure time is observed.

The first two approaches, the EC_{50} and LT_{50}, have been more widely applied. It is important to point out that there is a substantial difference between them: the latter provides, as a final outcome, the survival time of half of the treated organisms; when this time approaches infinity, the toxicity curve tends to an asymptote intercepting the concentration axis in a value called *incipient* LC_{50}, or *lethal threshold concentration*. In Figure 2.2, some unpublished experimental data on *Carassius auratus* and methylmercury are plotted to illustrate a case where this threshold concept shows its limitations. Effectively, there was a threshold for the first toxicity mechanism (probably a gill impairment), but continuing the experiment fish start dying again, according to another toxicity function with another asymptote (perhaps corresponding to a kidney or liver impairment). Data obtained by Abram (1967) with DDT in rainbow trout (*Salmo gairdneri*) seem to indicate the existence of a third asymptote (central nervous system lethal threshold?). The main problem with this approach is the identification of the number of possible lethal thresholds and the respective biological meaning. In addition to the problem of multiple thresholds, the results of LT_{50} measurements and relative spread parameters are expressed in units of time: this makes it difficult to apply the approach for regulatory purposes, where standards are currently espressed as concentrations or masses of contaminants.

In providing criteria for regulations the other method—the EC_{50} approach and its derivations—are the most commonly applied. *They indicate a concentration or a dose that is certainly dangerous, with related spread parameters.* By reducing this with an appropriate safety factor, it is possible to evaluate safe exposure levels. In the following sections, the EC_{50} approach is illustrated, for its historical and methodological significance and also for its still extensive application in regulations.

As mentioned before, the concentration corresponding to a 50% incidence of some adverse effect is called *median effective concentration*, EC_{50}. Similarly, ED_{50} indicates the *median effective dose*. The exposure time is short (e.g., 24 to 96 h) and, to express acute toxicity, the exposure time should also be indicated jointly with the median effective concentration: EC_{50}24h or LD_{50}48h, etc.

An example of a typical result when plotting *cumulative percent incidence* of the biological response (as mentioned, the directly observed incidence is cumulative) vs concentration is given in Figure 2.3. The effect, to be treated as percent incidence, must be *quantal* (meaning of the *all-or-none* type) such as death in lethality tests. With *continuous* effects, such as the increase or decrease in weight of treated plants or animals, these can be transformed into quantal data by setting a limit separating *effect* from *no effect* (e.g., below 10% variations = no effect).

Quantal data (enumeration data) are different from continuous variables (such as measures of concentration, weight, or length) due to their origin from classifications based on categories (e.g., male and female). Quantal variables, therefore, have only the possibility of varying by discrete units. To elaborate quantal measures, it is incorrect to apply statistical techniques based on a normal distribution of data, as these are distributed according to a binomial distribution. Statistical analysis of quantal data reflects initial approaches based on death as an endpoint, and as the most easily measured response (Bliss, 1935). Curves like that

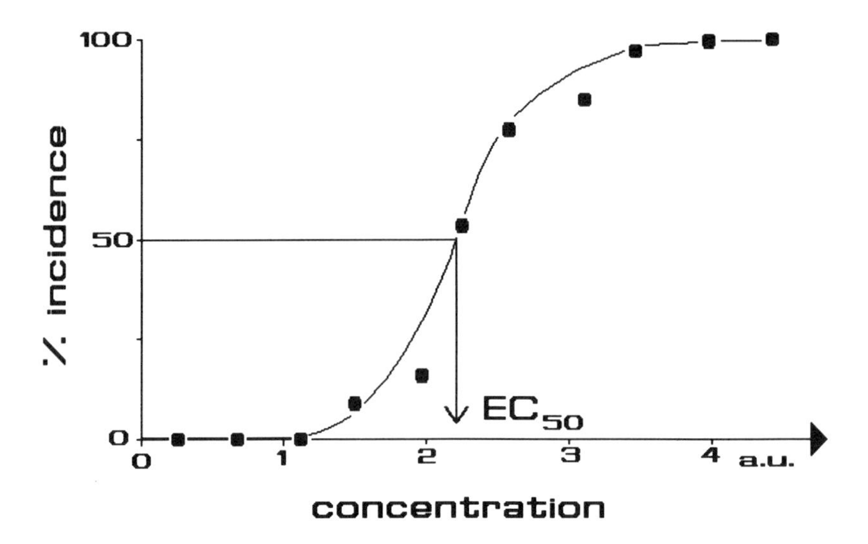

Figure 2.3. The classic toxicity curve % incidence (or biological response) *vs* concentration. a. u. = arbitrary units.

in Figure 2.3 are generally asymmetrical and nonlinear. Concerning this last point, however, when the elaboration is limited to the central part of the curve (i.e., from 10 to 90% effect incidence), results of different statistical approaches are not so dissimilar; differences become important when extrapolations outside the range of experimental data are carried out.

As far as the asymmetry is concerned, due to the reduced response rate at concentrations higher than EC_{50}, the logarithmic transformation of the abscissa may increase the symmetry of the curve. This can be done using base 10 (log), base *e* (ln), or other base logarithms: *the result is the same as all logarithms are directly proportional to each other* (e.g., log A = ln A/ln 10). The semilog plot will allow another modification of the curve to obtain a straight line: this can be done by the transformation of the response axis from percentages to probits (Bliss, 1935; Finney, 1971). An example of a typical log-probit plot is shown in Figure 2.4. The limitation of this approach is that the regions of the curve in the proximity of 0 and 100% incidence cannot be explored: 0 and 100% values must therefore be excluded from calculations. However, this approach is widely applied, particularly since the diffusion of personal computers. Manual calculations are long and tedious, but this problem has been overcome by means of suitable software such as that by Trevors (1987), which provides EC_{50} values, *95% confidence limits,* and *slope* of the log-probit curves. With the slope, the effective concentrations corresponding to probit 4 (or 15.9 % incidence) EC_{16}, and probit 6 (or 84.1 % incidence) EC_{84}, can also be obtained.

> *The 95% confidence limits define an interval such that, if all possible replicate 95% confidence intervals were determined in the same manner with test organisms from the same population, 95% of them would include the true EC_{50} of the population under the conditions of the test* (Stephan, 1977).

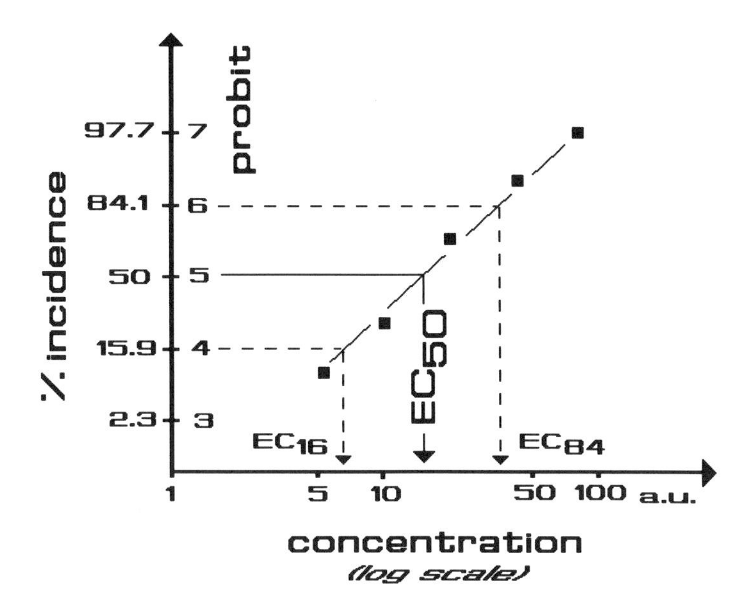

Figure 2.4. Log-probit plot of acute toxicity data. a. u. = arbitrary units.

The *slope of the log-probit correlations indicates the scattering of percent incidence data*, due to the extent of the individual variability in reacting to the toxic agent: in homogeneous populations of treated organisms, the slope tends to be higher than for more variable groups. This variability may be intrinsic in the examined group of animals or may be due to some scarce reproducibility of the experimental results. According to Hartung (1987), "the slopes of experimentally determined dose-response curves have no inherent relationship to the concept of *potency*, or *intrinsic toxicity,* or *relative toxicity*. These latter concepts, when used in conjunction with the slope of the dose-response relationship, are useful only when they are applied to specific mathematical dose-response models. Numerically, potency is a variable whose value usually changes as the dose regimen changes. It is constant only in special linear dose-response models", such as the linearized multistage model (see Chapter 2.2.2.4.2).

It is important to avoid confusion between the slope of the probit line, depending on the scattering of biological response data, and another slope, $S = C_{50}/C_{16} = C_{84}/C_{50}$, which refers to the scattering of concentrations C (where C_{50}, C_{16}, and C_{84} correspond, respectively, to probit 5, probit 4, and probit 6). This is the spread parameter of data log-normally distributed, having a geometric mean = median = C_{50} (Mackay and Paterson, 1984). This approach may be applied in the elaboration of a series of measures (such as those from a survey on the levels of contamination), when they are log-normally distributed. In this case, the cumulative frequency of the measures, in the ordinate axis, is transformed into probit (Calamari et al., 1991).

Going back to toxicology, a limitation of the probit approach is that *only one data set should be considered for the probit line estimation.* The EC_{50} values obtained for the same chemical, the same species, and following a standard approach may vary

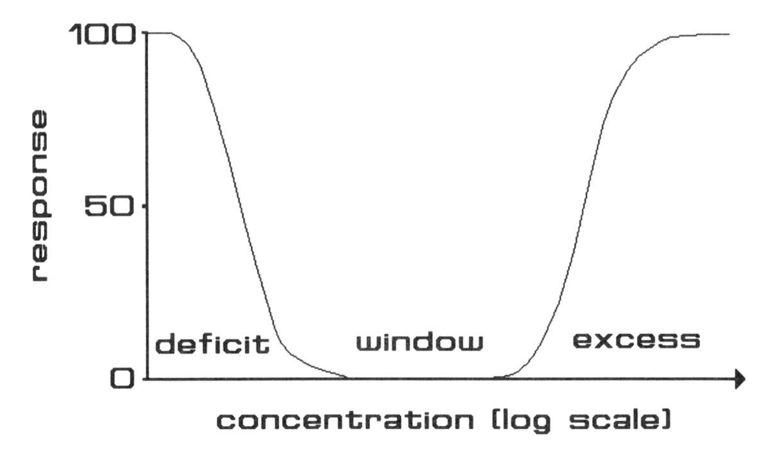

Figure 2.5. Anomalous toxicity curves: the case of an essential substance showing the *concentration window*.

considerably from one replicate to another within the same laboratory and between different laboratories. This variability is only partly due to technical, environmental, and biological factors: a significant contribution to the scattering of results is intrinsic to the nature of quantal variables (Lison, 1961). Thus, the combination of data from different experiments may reduce the error of the EC_{50} estimates. Unfortunately, this is not accomplished correctly by simply calculating the arithmetic mean of a series of observed responses for the same exposure level and then calculating the overall probit line. To overcome these difficulties, Hong et al. (1988) introduced the concept of the "grand probit line" by modifying the maximum likelihood probit method and incorporating a technique for parallel line probit analysis (Finney, 1971). The same authors developed a BASIC program for personal computers for estimating a best-fit probit line, with a single ED_{50} (or EC_{50}) value and relative confidence intervals, starting from multiple toxicity test data.

Anomalous Toxicity Curves: Differences in sensitivity distribution in treated organisms may lead to unusual toxicity curves. If, for example, in the treated group there are two subgroups characterized by different sensitivities to the toxicant, the resulting response-concentration or response-dose curve may appear as a double S-shaped function due to the bimodal frequency distribution of the biological responses.

Special toxicity curves are obtained in the case of essential chemicals (e.g., selenium): low and high doses, respectively, below and exceeding a certain dose range (*dose window*) may be toxic; while in the dose (or concentration) window, the substance may not be toxic at all (Figure 2.5).

Three-dimensional Approaches: Both the duration of exposure and concentration (or dose), influence the biological response. Three-dimensional models have been developed from the *response-concentration-duration surface*, particularly with carcinogen substances (Hartley and Sielken, 1977), for the calculation of the ED_{01} (effective dose 1%).

2.2.2.2. Chronic Toxicity

In chronic toxicity studies, the traditional goal is to calculate *thresholds*, or those levels of exposure to toxicants which are not able to induce any detectable adverse effect in the treated organisms. The concept of a toxicity threshold is not correctly applicable to those substances directly interacting with DNA, modifying its code, though it is possible to imagine a threshold in the case of, for example, an enzyme inhibition where repair mechanisms may be effective below a given exposure level (the toxicity threshold). The *chronic toxicity threshold* is obtained by means of chronic studies where the exposure levels are lower than for acute tests, and the exposure time is longer (generally corresponding to one or more reproductive life cycles). Currently, the *toxicological endpoint* is not lethality, but more subtle effects such as the reproductive success. Generally, different sublethal effects are jointly considered to point out the threshold, which is the limit between observed-effect and no-observed-effect exposure levels. The *no-observed-effect level* (NOEL) produces an approximation, by defect, to the chronic threshold region. In practice, the problems arise from the difficulty in distinguishing these effects from the background noise (are they effects or not?), with the possibility of both false positive and false negative errors. The former refer to exposure levels which are not effective but are considered as producing some effect; the latter occur when effective exposures are taken as NOEL.

Test organisms may vary according to the goal of the experiment. For example, if the aim is to evaluate the toxicity of a substance to humans, a mammal is selected (e.g., rat, mouse, guinea pig, cat, dog, or non-human primate); in evaluating effects in aquatic environments, fish, crustaceans, mollusks, and algae are selected. A toxicological evaluation should be based on long-term effects on more that one species as representative of the system under study. When possible, research should be carried out using standard models—standard experimental procedures and standard organism strains. Genetic factors may be significant, and the use of a different strain may lead to the introduction of an unknown variability factor in both effects and responses.

Duration of the treatment will depend on the species: for the water flea *Daphnia magna*, 21 d are enough; whereas, it will be longer for fish such as the fathead minnow *Pimephales promelas* or rainbow trout *Salmo gairdneri*, which need to be observed during growth, development, and production of the first generation. As an alternative to the life cycle test, *early life stage tests* are conducted. In the case of fish, these are exposed for about 1 month from the fertilized egg stage through that of embryo, larva, and juvenile development, so as to obtain an indication of the long-term effect over the entire the life cycle, assuming that the early life stage is the most sensitive.

Toxicological endpoints concern the effects that are taken into consideration. These include growth rate, development, and reproductive success, but more refined endpoints may be selected. Endpoints have been grouped as *biochemical and physiological,* and *behavioral and hystological* (Rand and Petrocelli, 1985).

The observed effects in *treated* animals and related measures at various concentrations or doses are compared with results in *controls* (i.e., treated at dose "zero"). Statistical techniques are available to point out when the effect in one or more of the treated groups is significantly different from that of the control group. Effect concentrations are those with these effects, while no-effect concentrations are those where these effects are not found or are not statistically significant *(no-observed-effect concentration, NOEC*, analogous to the *no-observed-effect dose*,

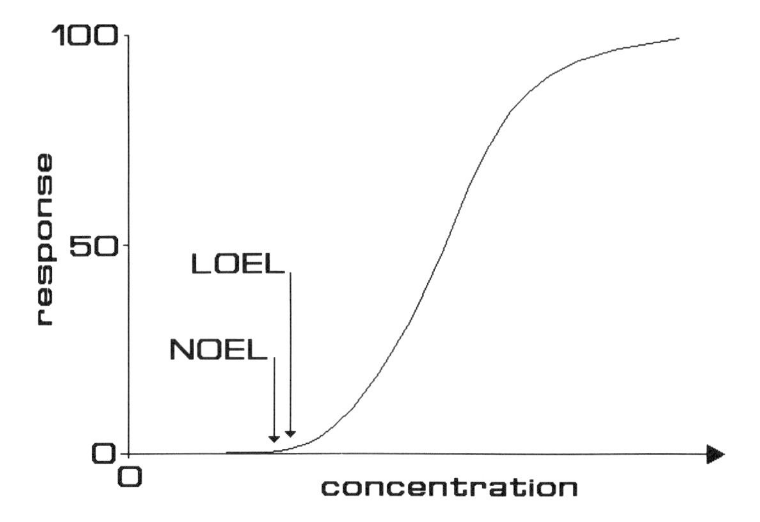

Figure 2.6. Schematic representation of NOEL and LOEL concepts.

NOED, or to the more general *no-observed-effect level, NOEL*). The *lowest-observed-effect level, LOEL,* is the lowest exposure level at which significative effects are obtained (Figure 2.6). The threshold is located in the middle of NOEL and LOEL, and indicates the separation of effect from no-effect concentrations; it is generally is calculated by taking the geometrical mean of LOEL an NOEL.

From the previous discussion, it can be observed that the *effect concentration deals with the toxicological effect,* while the *effective concentration,* applied to acute toxicity measurements, *refers to the biological response.*

A key point in the exploration of the threshold region is the statistical significance. This is expressed by a P-value = 0.05, indicating the probability that a random event caused the observed difference in the effect. Thus, every effect difference with a P value greater than 0.05 may be correlated with the exposure to the toxicant. The P = 0.05 value indicates the statistical significance level (also indicated by 95%, or the complement to unit of P as percent, indicating that the observed difference has a 95% probability of not being due to a random fluctuation).

Usual statistical techniques applied to chronic toxicity experimental data include two steps:

- Analysis of variance
- Multiple comparison test

In the case that a solvent (*carrier*) is applied to increase the solubility of the toxic in the medium (e.g., water in aquatic toxicological tests), an additional experimental control with the carrier is needed to verify whether the carrier alone produces any toxicity. In the case of significant effects with respect to the controls, the control with the carrier will be used as the new control group.

When chronic toxicological endpoints are expressed in continuous variables, the analysis of variance between treated and controls can be directly carried out.

However, if data are expressed in quantal variables (i.e., as a proportion, p, of test animal responding to the exposure) before applying suitable statistical techniques for the analysis of variance, all p values must be transformed to homogenize the variance. In proportions as p values are, variance changes as a function of the value of p: around 0 and 1 (i.e., 0 and 100%) it is minimal, and at $p = 0.5$ (50%) reaches the maximum (Lison, 1961). When proportions are derived from the same number of treated individuals, the variance may be homogenized by the angular transformation of p values to arc-sine $(p)^{1/2}$:

$$p \rightarrow \text{arc} - \text{sine}(p)^{1/2} \rightarrow \text{analysis of variance}$$

After the arc-sine square root of p transformation and the analysis of variance by means of a multiple comparison test such as the Williams' test (Williams, 1972), the result will be an indication of *statistically significant* or *not statistically significant* effects due to the exposure to the toxicant. The lowest concentration producing a statistically significant effect, LOEC (*lowest-observed-effect concentration*), and the highest concentration not producing the same effect, NOEC, are obtained. As mentioned before, the estimated chronic toxicity threshold level is located in the region limited by NOEC and LOEC (or NOEL and LOEL), and indicates the limit below which the effect corresponding to the most sensitive of selected endpoints is not expected.

In tests with fish, the threshold concentration has been also called *maximum acceptable toxicant concentration*, MATC (Mount and Stephan, 1967).

The high cost of long-term toxicity experiments was probably the reason for attempts to correlate acute toxicity and chronic toxicity data. Mount and Stephan (1967) proposed the use of an *application factor*, AF.

$$AF = MATC / LC_{50} \tag{2.1}$$

If the MATC is not known, but NOEC, LOEC, and LC_{50} are known, the AF will be in the interval NOEC/LC_{50} and LOEC/LC_{50}. Once AF is obtained for one species, assuming that it is dependent more on the properties of the toxicant than on the properties of the biological system selected for the test, it is possible to apply the AF to calculate the MATC for other species starting from acute toxicity values (MATC = AF LC_{50}).

In the second half of the 1970s, the use of AF was replaced by the early-life-stage and by the most-sensitive-stage tests. However, in recent years, these have been criticized. Experiments on early-life stage and on most-sensitive-stage do not seem able to substitute the life cycle tests since reproduction is the weakest point of the cycle of a species. According to Suter et al. (1987), the most sensitive effects result in the reduction of the *reproductive success* of the exposed populations.

To extend experimental data from the observed population to a larger field population of the same species and then to other species and to biological communities, arbitrary security factors ranging from 1/100 to 1/10 are applied to the

MATC values (or to other espressions of the chronic toxicity threshold) in order to estimate environmentally *safe levels*; that is, those levels of exposure which are supposed not to produce deleterious effects. In a similar way, regulatory agencies may arbitrarily apply a safety factor to a NOEL to determine acceptable doses of contaminants to humans, such as the *acceptable daily intake,* ADI. Notwithstanding their intrinsic limitations (see Chapter 2.2.2.3.), which suggest the need for further investigations to develop improved tools, the indications from NOELs and MATCs continue to be largely applied in hazard assessment of nonmutagenic environmental chemicals.

2.2.2.3. Experimental Design and False Negative Errors

Refinement of current methods for safe levels evaluation is a field where active biological and statistical research is in progress. One of the main problems in current approaches is the use of NOEL (the no-*observed*-effect level) as it was a *no-effect level*, NEL, without taking into account that NOEL is simply an exposure level at which the difference in responses between treated and control groups is not statistically significant (or the *null hypothesis*, H_0, is not rejected). This does not necessarily imply that effects were not occurring. A related key point is the optimization of the number of test organisms (*sample size*) for detecting specified *biological effect sizes* by means of suitable statistical approaches. Animal welfare activists claim that too many animals are afflicted, or are killed, to measure toxicity (Ecobichon, 1992). On the other hand, *when the number of tested organisms is too small, there is the possibility of drastically overestimating safe exposure levels.* The NOEL is dependent on the probabilities that differences between the treated organisms and the control group will be significant. In statistical terms, this indicates that the NOEL is directly related to the *power* (i.e., *statistical power*) of consecutive tests of the null hypothesis (no difference between treatment and control groups).

Recently, Oris and Bailer (1993) have analyzed the case of *Ceriodaphnia dubia* survival and reproduction toxicity tests, a standard method for estimating chronic toxicity of effluents, and receiving waters, to freshwater organisms (Weber et al., 1989). The standard test starts with <12-h-old neonates and measures the survival and reproductive output of 10 individuals treated with five different concentrations of toxicant (plus a control group) over a 7-d period. At 25°C, this period is sufficient for the growth and maturation of the animals and for the production of three broods of young. According to the assumptions made for the distribution and variance of data, parametric or nonparametric tests are applied. The most common parametric statistical techniques for data normally distributed are the *t*-test with Bonferroni modification or the Dunnett's test. *The NOEL is the highest concentration of tested substance for which the null hypothesis, H_0, is not rejected, and the LOEC is the lowest concentration for which H_0 is rejected.* A presumably safe concentration, the *chronic value*, ChV (another expression for the chronic toxicity threshold), is calculated as the geometric mean between NOEL and LOEL. The key role of NOEL values suggests that the probability of *false negative* (i.e., NOELs which are effect levels) should be considered in experimental planning. Calling α the probability of false positive (rejecting H_0 when true), β the probability of false negative (accepting H_0 when not true), P the *statistical power* of the test, where $P = 1\text{-}\beta$, and c a proportionality constant to calculate the variance in the treated group from the variance of controls, Oris and Bailer (1993) have shown that: starting from 53 independent *Ceriodaphnia* survival and reproduction toxicity tests and combining different values of α, statistical power, and variance ratios c, the standard sample size of 10 organisms is capable of detecting a range of 31 to 100% reproductive inhibition. This wide range of minimum detectable inhibition values indicates that the ability to avoid false negative with current standard sample sizes is quite poor. The same authors propose that standard tests should be established on the basis of their ability to detect a specified level of reproductive inhibition and suggest that "the question of biological significance must be revisited. Indeed, given the power of the toxicity test with the current sample size, one cannot assume that statistical significance will reflect biological significance."

There is an additional problem in the current NOEL approach, even worse than the limitation discussed heretofore: statistical power increases with increasing sample size and by reducing the variability inside the tested groups (Crump, 1984). In other words, this means that an increase in number of test organisms and an increase in the homogeneity of experimental design will reduce the value of NOEL, demonstrating that the previous, more rough NOEL was an effect level. This intrinsic characteristic of NOEL shows its main limitation for ecotoxicological applications: *rough approaches overestimate safe levels.*

Probably, after some refining, methods starting from *effect levels* [such as the concentration corresponding to the 1% response, EC_{01}, and producing estimation of concentrations corresponding to an *acceptably small effect* by a linear estrapolation of EC_{01} to lower response levels (Chen and Kodell, 1989)], are more suitable than the NOEL approach. In this way, the problem is reversed: an increase in the sample size and in the precision of experimental design will lead to *an increase of the acceptable small effect level.* Recently, Hoekstra and van Ewijk (1993) have proposed, as an alternative for the NOEL, a two-step procedure that involves finding the dose whose response is at most 25%, followed by linear extrapolation to a dose whose response is acceptably small. New approaches to the problem of evaluating toxicity thresholds appears to be more suitable in preventing environmental damage.

2.2.2.4. Mutagenic and Carcinogenic Substances

Methods discussed above lead to an estimation of a *threshold region* in which a toxicity threshold is located. Consequently, below the threshold, no effects are observed and, taking into account experimental error and error due to *statistical inference* (from one part to the entire group and from one to several species), environmental no-effect levels are derived. This may be appropriate, with the limitations discussed in the previous paragraph, when there are some biological reasons to support the assumption of the existence of a threshold exposure level.

These reasons are based on one or more of the following mechanisms (Pfitzer and Vouk, 1986):

- Homeostatic, defense, and/or repair mechanisms are able to counterbalance the effect at the threshold exposure level, or below
- Absorption, transport, metabolism, and elimination processes do not allow for a sufficient dose to reach the biological target
- Eventual effects smaller than the effect considered as toxicological endpoint in the chronic test will not subsequently progress

When one single molecule can interact with a biological system initiating a chain of events that may produce a toxic effect, the threshold approach is not appropriate. This is the case for mutagenic and carcinogenic substances. Mutations in germ cells are an essential life mechanism, providing the instrument for evolution. When mutations occur in somatic cells, they may produce tumors.

Many chemicals have been shown to produce detectable levels of mutagenic and carcinogenic effects. For these substances, it is still correct to measure acute and chronic toxicity; however, it is not enough, due to the difficulty of assuming a toxicity threshold higher than zero.

The practical problem is to produce a response-dose relationship able to quantitatively correlate the exposure to the number of cancer induced, to say "with this exposure level, the added cancer incidence will be, for instance, 1/100,000, or 10^{-5}, after a lifetime exposure". From this statement, one can see the actual difficulty in carrying out such an experiment. Clearly, it is not possible to expose, under controlled conditions, 1,000,000 rats for 3 years (the rat life span) in order to observe 10 new cancers in a natural occurrence of 50 or more cancers of the same type. There are two alternatives:

- Take microbes or single mammalian cells as models: short life-span and little volume
- Carry out tests with rodents, but at very high doses, to increase the incidence of cancer, thereby making it easier to distinguish the effect in treated animals with respect to controls; the problem will be how to extrapolate observed effect-dose relationships down to low exposure levels

2.2.2.4.1. Short-Term Tests (in vitro *Tests) and Biomarkers.* One of the more frequently used tests for mutagenesis is the Ames test (Ames, 1971) which uses the bacterium *Salmonella typhimurium*, normal strains of which have the capability to synthesize the amino acid histidine and can be grown in culture media without histidine. Certain *Salmonella* mutants have lost this possibility and are not able to develop in a histidine-free medium. When these mutants are placed in histidine-free media together with a mutagen chemical, it could happen that a *reverse mutation* will allow for growth of bacteria. Several different mutant strains of *Salmonella* can be used to measure different mutations. The limitation of this approach is that it is only possible to measure a few well-defined mutations and not all the possible mutations. Another disadvantage, when applying the Ames test as a model for mammals, lies in the use of a procaryotic genoma which does not possess mammalian DNA repair mechanisms (Flamm and Scheuplein, 1988). Finally, the differences in the possibility of a metabolic activation between *Salmonella* and other species have to be considered, particularly for those mutagens or carcinogens for which this activation is required. Several carginogens, called *procarcinogens*, need to be biotransformed to highly reactive radicals, which are the actual active substances. To overcome this limitation, homogenates of mammalian liver, extremely rich in chemical-activating enzymes [e.g., the mixed function oxidase (MFO) system] are added to the incubation mixture (Ecobichon, 1992).

Another *in vitro* approach is based on mammalian cell cultures, from a variety of human tissues (DeMars, 1974). In these tests, *forward mutations* are detected, involving the use of purine and pyrimidine salvage pathways. If I-thioguanine is used to poison normal wild cells able to metabolize it to ribosyl phosphate, this can be incorporated in the DNA of the cell, thus causing death. Mutant cells, not able to carry out this transformation, will not be able to poison their DNA and will survive. Forward mutations appear to have the possibility of reacting to a series of events wider than the reverse mutations described above.

Laboratory methods have been developed to measure the actual amount of chemical interacting with a molecular target (DNA, RNA, or protein), or the *biologically effective dose* (Perera, 1988). These allow for early identification of cancer hazard and produce estimates of the potential risk on the exposed organisms.

The measurable alterations of normal biochemical or molecular processes due to an effective dose of pollutant are called biological markers or "*biomarkers*". For carcinogens, these include: (1) carcinogen-DNA and (2) carcinogen-protein *adducts. Adducts are the result of covalent binding of an electrophilic group of the carcinogen and DNA, RNA, or a protein.* Protein adducts are related to DNA adducts; the modified DNA may lead to gene mutation (*cancer initiation*) and cancer (through *promotion* and *progression*).

To measure the concentration of adducts in treated biological materials, several methods based on different approaches, including biochemical and immunological techniques, are available. Cytogenetic methods are aimed at estimating chromosomal aberrations, sister chromatid exchanges (SCE), and micronuclei (MN) in lymphocytes (Perera, 1988). The advantages of biomarkers are (Lutz, 1979):

- Low-dose sensitivity (no extrapolation from high to low-dose is needed)
- Possibility of directly testing the species concerned
- Possibility of investigating intraspecific variability
- Good correlation between the quantity of adducts and carcinogenic potency

Many cancers may be caused by DNA damage, followed by genetic changes that may produce a malignant cell, often many years after the initial exposure.

Carcinogenesis evolves through three different phases: *initiation, promotion,* and *progression.* Chemicals that are active mutagens and carcinogens are either polar, or generate—through the metabolic activity of organisms—polar compounds. These chemicals attack the nucleophilic center in nucleic acids and proteins, resulting in the formation of *covalent adducts. The DNA adducts are considered to represent the initiating event that may lead to mutation and malignant transformation.* A cancer promoter is defined as an agent that results in an increase in cancer induction after long-term application subsequent to treatment of an animal with an initiator. At a particular dose level and treatment schedule, a carcinogen that does not require supplementary promoter activity is termed *complete.* The progression of a neoplastic cell into a malignant tumor is the last stage of cancer formation.

From a quantitative point of view, carcinogenic potency is proportional to the ability to form covalent bonds with DNA in a large number of substances (Lutz, 1979). Covalent DNA bonds may vary in stability: some of them are very unstable and are spontaneously removed; others need an enzymatic-mediated process, which is the DNA *repair* mechanism. These mechanisms are relatively similar in mammals and in bacteria, indicating that these systems have been conserved during the course of evolution.

The measurement of DNA adducts or related compounds in mammal urine is applied to measure effects of recent exposure to genotoxic chemicals, and to the effectiveness of DNA repair mechanisms. The measurement of the concentration of more stable DNA adducts is made in accessible animal and human cells (e.g., white blood cells) by means of physicochemical and radiochemical approaches. Recently, immunological techniques have been introduced. In particular, monoclonal antibodies have been applied to bind carcinogen-DNA adducts (Müller and Rajewsky, 1981).

Protein adducts are more stable than DNA adducts; their concentration can be detected in hemoglobin (from red blood cells) to obtain an evaluation of the effects of an exposure integrated over the lifetime of the protein (Ehrenberg and Osterman-Golkar, 1980).

Although there are some exceptions (i.e., chemicals which are carcinogens but do not produce any adduct), in the majority of cases it was observed that, at low doses, adduct formation is directly proportional to the dose (first-order kinetics). This indicates that there is no toxicity threshold for mutagens: very low exposure levels are able to induce the production of significant levels of adducts. Correlations between DNA adducts, and the frequency of induced mutations, may be applied in assessment procedures.

These short-term tests made it possible to test a great number of chemicals with encouraging results: tests on polynuclear aromatic hydrocarbons (PAH) and aromatic amines indicated the high mutagenic potential of some of these compounds, in accordance with the findings in cancerogenesis tests with laboratory animals. However, as reported by Flamm and Scheuplein (1988), a "growing number of nonmutagenic compounds are capable of inducing cancer in the test animals, rats, and mice". Although the overlap between tests in *vitro* and *in vivo* is not perfect, the information from these rapid and low-cost, short-term tests remains an essential tool in improving toxicological knowledge.

2.2.2.4.2. Animal Bioassay Approach. These are mainly carried out to provide effect-dose relationships for carcinogenic substances in order to estimate the additional cancer incidence in a given population after lifetime exposure as a function of the exposure level. In general, these data are produced to evaluate the cancer risk in humans; consequently, epidemiological data, when available and sufficiently reliable, is also used for comparison with laboratory-model findings. Cancer incidence in wild animals does not seem a priority. In any case, it is obvious that *if animal models are developed to simulate man, the same information could be adapted and applied to wild animals.*

As already mentioned, the main problem with animal bioassay data is the extrapolation of the obtained effect-dose curve down to very low exposure levels, given that these low-level effects are not directly measurable. Before applying any mathematical approach, an understanding of the process involved in carcinogenesis is necessary. The current theory is that most of the cancer-producing chemicals are also able to produce irreversible damage to DNA, as data on mutagenesis and carcinogenesis seem to indicate. If this is the case, *the discontinuous quantal and nonlinear responses, such as the incidence of cancer, must be related to a linear nonthreshold effect-dose relationship.* Among the possible approaches, the *linear nonthreshold model* (or the *linearized multistage model*) represents the best, albeit limited, scientific basis for any of the current extrapolation mathematical approaches. Furthermore, this model tends to be conservative, in the sense that it overestimates the risk; therefore, safety is probably guaranteed, despite a high degree of uncertainty.

The linearized multistage model is derived from the multistage one, introduced by Armitage and Doll (1961) and modified by Crump (1981). The original multistage model is expressed as follows.

$$P_{(d)} = 1 - \exp\left[-\left(q_0 + q_1 d + q_2 d^2 + \ldots + q_k d^k\right)\right] \qquad (2.2)$$

where $P_{(d)}$ is the lifetime probability of cancer (*lifetime cancer risk*) at dose d, q > 0, and i = 1,2...k. The extra risk over background rate at dose d, $A_{(d)}$, is:

$$A_{(d)} = 1 - \exp\left[-\left(q_1 d + q_2 d^2 + \ldots + q_k d^k\right)\right] \qquad (2.3)$$

where

$$A_{(d)} = \left[P_{(d)} - P_{(o)}\right] / \left[1 - P_{(o)}\right] \qquad (2.4)$$

with $P_{(o)}$ indicates the background cancer risk.

The extra risk function is obtained, leading to the $A_{(d)}$ values corresponding to a given dose with 95% upper confidence limits. Special software has been developed by Crump and Watson (1979). The 95% upper confidence limits on the extra risk $A_{(d)}$ and lower confidence limits on the dose producing a given risk are determined from a 95% confidence limit, q_1^*, on parameter q_1. For $q_1 > 0$, at low doses:

$$A_{(d)} = q_1 \, d \qquad (2.5)$$

The product q_1^* d is a 95% upper confidence limit on the extra risk, and R/q_1^* is a 95% lower confidence limit on the dose producing an extra risk R.

The 95% upper confidence limit is always linear and the slope q_1^* is taken as an upper bound of the potency of the chemical in inducing cancer at low doses (Anderson, 1985). The slope q_1^* is also called the *upper bound slope.*

The upper bound slope may be adjusted for exposure duration and species differences to give an estimation of carcinogenic potency in humans, B_H (human upper bound slope) (Stara et al., 1987).

$$B_H = \left[q_1^* \left(70/W_A\right)^{1/3}\right] / \left[\left(l_e/L_e\right)/\left(L_e/L\right)^3\right] \qquad (2.6)$$

where q_1^* = upper bound slope (kg d/mg)
70 = standard human body weight (kg)
W_A = body weight of the test animal (kg)
l_e = length of exposure (d)
L_e = length of observation period in animal experiments (d)
L = lifespan of the animal (d)
B_H has the same dimensions as the upper bound slope: a time multiplied by a reciprocal of a dose, and units of d·kg/mg.

From q_1^*, the value of B_H can be calculated. A list of some B_H values is given in Table 2.1.

The knowledge of B_H allows calculation of the *intake rate, I, in mg/d, associated with a selected human lifetime cancer risk*:

$$I = (70 \times R) / B_H \qquad (2.7)$$

where R is the lifetime risk (e.g., 10^{-5}). The equation can also be applied in the inverse form:

$$R = I \, B_H / 70 \qquad (2.8)$$

to obtain the risk, R, associated with a given toxicant intake rate *I*.

Table 2.1. Human Upper Bound Slope Values, B_H, for Some Chemicals

Chemical	B_H (kg·d/mg)
Aldrin	11.44
Arsenic	14.00
Benzene	0.052
Benzidine	234.13
Carbon tetrachloride	0.083
Chloroform	0.18
Hexachlorobenzene	1.68
DDT	8.42
2,3,7,8-Tetrachlorodioxin	425,000
Hexachlorocyclohexane:	
Technical grade	4.75
Alpha isomer	11.12
Beta isomer	1.84
Gamma isomer	1.33
PAH	11.53
PCBs	4.43
Trichloroethylene	0.013
Vinyl chloride	0.017

From Anderson, 1985. With permission.

Thus, if the human upper bound slope, B_H, is 3.5 kg d/mg, and the intake rate, I, is 10^{-3} mg/d, the associated risk, R, is 5×10^{-5}.

The approach illustrated here, despite its intrinsic limitations, represents a significant improvement on previous qualitative evaluations, and is probably the best currently applied to carcinogenic risk assessment. At the moment, its application is limited to humans; however, it could be applied to other animals and represents a call for research. Besides, it is important to remember again that by increasing statistical power, with increasing sample size and by reducing the variability inside the tested groups, the upper bound slope will decrease (Crump, 1984). In other words, this means that the upper bound slope tends to overestimate cancer risk.

2.3. TOXICOLOGY OF MIXTURES

Everything discussed in Section 2 heretofore clearly demostrates the deep links and common roots between environmental toxicology and classic toxicology. Although the animal model has changed (e.g., from rat to Daphnia and the endpoint is *immobilization* rather than death), the differences lie more in laboratory techniques than in the scientific approach. One of the major differences between man and other animals is that the former may be exposed to hazardous chemicals at the workplace, at home, driving cars, and in other artificial environments. Man-made environments tend to be increasingly similar and standardized, with the possibility that, from a certain point of view, there is the risk that toxicological data on humans becomes more reliable than data from laboratory animals (are we exchanging roles?).

When chemicals are released into the environment, if not already produced as mixtures, they immediately become mixtures.

The assessment of the effects of chemical mixture may follow two approaches:

- Testing mixtures
- Use of toxicity data on single chemicals

2.3.1. Direct Tests

In studies for testing mixtures, these are considered as a single pure substance and the methods are similar to those for pure chemicals. In these studies, mixtures of pure chemicals and various environmental or man-made materials have been tested; these include drinking water (after chlorination), airborne particles, cigarette smoke, automobile exhaust, fly ash from municipal waste incinerators, industrial liquid effluents, and municipal wastewaters.

When the mixture composition is known, it is interesting to compare its toxicity with that of its pure components. This approach may lead to the development of methods for evaluating the mixture toxic potential, starting from that of it components.

2.3.2. Use of Toxicity Data on Single Chemicals

The approach of combining toxicity data of components of the mixture to the mixture itself depends on the type of *joint action* involved. According to Bliss (1939), this can be: *independent, similar, synergistic, or antagonistic*. In the first case, the chemicals have different modes of action and act independently, and the toxicity of the mixture can be predicted by the effects of single components. In the second case, the components produce similar but independent effects (i.e., one component may be substituted by a constant proportion of another). In the last two cases, the toxicity of the mixture depends on the proportion of the components (i.e., the presence of one may increase or reduce the toxicity of others as a function of the relative proportions). Hoel (1987) recently reviewed the statistical research carried out to establish the combined action of two or more chemicals in a mixture.

In the absence of known different mechanisms, the similar joint action approach or *similar additive action* is the most widely used for preliminary screening and for regulatory purposes (Alabaster et al., 1988). An example of a mixture of two chemicals may be useful to illustrate the subject. According to Finney (1971), if the log-probit plots are assumed to be linear, the reciprocal of the predicted LD_{50} for the mixture, $1/LD_{50}mix$, can be calculated as follows:

$$1/LD_{50}\text{mix} = p_A/LD_{50}A + p_B/LD_{50}B \quad (2.9)$$

where p_A and p_B are the respective proportions of the components A and B.

If, for instance, 10g of an AB mixture contain 3g A and 7g B ($p_A = 0.3$ and $p_B = 0.7$) and LD_{50} of A and B are known, from Equation 2.9, the LD_{50} of the mixture is easily obtained. However, this approach is based on the assumption of an additive action of the components. To verify this assumption data from toxicity tests with A, B, and their mixture (A + B) are necessary to solve the following equation: $xTU_A + yTU_B = 1TU_{(A+B)}$ where TUs are the "toxic units" by Sprague and Ramsay (1965) indicating the fraction a concentration is in respect to the LD_{50} for a given biological species. The type of joint action depends on the values of x and y which satisfy the equation: when $x + y > 1$ the action is less than additive; when $x + y < 1$ the action is more than additive; when $x + y = 1$ the joint action of A and B is additive (Lloyd, 1987).

Lloyd (1961) found additive action for copper and zinc sulfate mixtures in rainbow trout (*Salmo gairdneri*) in both soft and hard water. Sprague and Ramsay (1965) found similar results with mixtures of copper and zinc salts in juvenile Atlantic salmon. Metal and surfactant mixtures were found to show additive toxicity, or a little more than additive effects, in rainbow trout (Calamari and Marchetti, 1973). Smyth et al. (1969) carried out an experimental study with 27 industrial solvents in rats by pairwise testing for LD_{50}: the 350 pairs of tests showed that a large proportion was following the additive action model. There were deviating pairs on both sides of the central values, demonstrating the reliability of the approach.

The relationship in Equation 2.9 can be is applied to response levels other than the 50% (e.g., EC_{01}) and to more than two chemicals. The general form is as follows.

$$1/EC_x mix = \sum_{i=1}^{n} \left(\mathbf{p}_i / EC_x \text{ for chemical } i\right) \qquad (2.10)$$

where p_i indicates the fraction of chemical *i* in the mixture, and x is the proportion responding.

When two or more hazardous substances with similar and additive toxic action are present in a medium, there is the need to evaluate whether or not their combined effects exceed *environmental quality criteria* or, from a legal point of view, *environmental standards* (i.e., legal threshold values, such as the *threshold limit value*, TLV). In the absence of information to the contrary, the additive model is applied to calculate the following sum:

$$\sum_{i=1}^{n} = \left(C_1 / T_1 + C_2 / T_2 + C_3 / T_3 + \ldots C_n / T_n\right) \qquad (2.11)$$

where C is the concentration in the medium, and T the threshold limit value, TLV, for each chemical. If $\sum_{i=1}^{n}(C_i / T_i)$ is equal or less than 1, then the threshold limit is not exceeded; when $\sum_{i=1}^{n}(C_i / T_i) > 1$, the limit is exceeded.

Table 2.2. Examples of Calculation of Compliance of Measured Concentrations in Air and the *Threshold Limit Value*, TLV, for Two Hypothetical Solvent Mixtures

	Solvent		
	A	B	C
Conc. in air C (mg/m³)	5	20	2
TLV (mg/m³)	10	100	4
C/TLV	0.5	0.2	0.5
		Σ(C/TLV) = 1.2, limit exceeded	
Conc. in air C (mg/m³)	4	10	0.8
TLV (mg/m³)	10	100	4
C/TLV	0.4	0.1	0.2
		Σ(C/TLV) = 0.7, limit not exceeded	

These approaches are currently applied in toxicology and industrial hygiene (Plunkett, 1976). An example of calculation is reported in Table 2.2. It is important to remember that in the case of independent action, each component must not exceed its own threshold limit value, such as the TLV. For example, if the air in a workplace contains 0.1 mg/m³ of lead (TLV = 0.15 mg/m³) and 0.7 mg/m³ of sulfuric acid (TLV = 1 mg/m³), we have 0.1/0.15 = 0.67 for lead, and 0.7/1 = 0.7 for sulfuric acid; therefore, the limits are not exceeded.

The mutagenicity of complex mixtures has also been investigated using a similar approach (i.e., comparing the toxicity of the mixture and of the components). At present, due to the complexity of possible interactions, there are no guidelines for predicting the mutagenicity of a mixture of two or more components, even if the mutagenicity of each single compound is known. It may happen, for instance, that two components, neither of which is mutagenic in the Ames test, give rise to a mutagenic mixture (*comutagenesis*; Wakabayashi et al., 1982).

2.4. FIELD STUDIES: *BIOLOGICAL MONITORING* AND *NONDESTRUCTIVE BIOMARKERS*

Biological monitoring in the field deals with two main objectives:

- The measurement of "internal-dose" or "internal exposure level" of selected organisms
- The measurement of effects

As far as the first objective is concerned, measurements of chemical contaminants and relative metabolites are carried out. This should be coupled with the second objective: to test possible cause-effect relationships. As previously mentioned, biological markers, or biomarkers, are measurable alterations of normal biochemical or molecular processes due to an effective dose of pollutant. When the alterations are measurable and are related in a quantitative way to the exposure to a known chemical, the information from biological markers may be significant in hazard assessment.

Current applications include the measure of the degree of inhibition by inorganic lead of the erythrocyte enzyme δ–aminolevulinic acid dehydratase (ALA-D). This enzymatic activity inhibition has been observed in mammals and teleost fish (Hodson et al., 1984) and may be applied to detect damage due to lead exposure.

In animals exposed to some heavy metals (e.g., Cd, Cu, and Zn), there is an induction of cytosolic, low-molecular-weight proteins: metallothioneins (MTs). These are involved in the regulation of intracellular levels of different trace elements and may act as detoxifying agents by binding toxic cations.

Another biomarker is the hepatic microsomal cytochrome P450 monooxygenase system (also called "mixed function oxidase system," MFO). This can be induced by polycyclic aromatic hydrocarbons (PAHs), 2,3,7,8 tetrachlorodibenzo-*p*-dioxin (TCDD), coplanar polychlorinated biphenyls (coplanar PCBs), and other structurally similar compounds. The MFO activity can be measured by means of different substrata. In the case of fish, the response of MFO activity is almost limited to PAH-type inducers and can be considered more specific than in other classes of vertebrates. In fish the ethoxyresorufin-O-deethylase (EROD) and the aryl hydrocarbon hydroxylase (AHH) activities are the most sensitive to PAH induction (Haux and Förlin, 1989).

Other alterations, such as the inhibition of blood esterases by organophosphate insecticides, DNA-adduct formation, etc., may be applied to wild animals to obtain an indication of the pollution status in a given environment.

One of the limitations of field studies is that although biomarkers may be effective in detecting the occurrence and the significance of adverse effects, they are often not highly specific and may be generated by different pollutants. A major problem is the difficulty of knowing exposure levels and/or exposure duration.

The study of biomarkers implicates the "collection" of animals, or organs or tissues, with the consequent decease of the animals. To avoid the "disaster" for endangered species of an extensive "sampling", nondestructive approaches have recently been developed (*nondestructive biomarkers*; Fossi and Leonzio, 1993). Using these methods, the required biological material (e.g., blood, milk, hair, or feathers) is selected in such a way that it can be taken without damage to the animals, as for humans.

An interesting approach was introduced by Ramaiah and Chandramohan (1993): these authors found out how bacterial luminescence, very sensitive to pollutants, is affected in marine coastal environments receiving various types of effluents. In a 7-year study along the coasts of India it was seen that luminous colonies were abundant (>10 and often 15% of the total colony forming units) in pollution free areas, while heavily polluted sites were characterized by no luminous colonies. Luminous bacteria were identified as *Photobacterium leiognathi, Vibrio harvegi, V. fisheri* and were similar in number and species composition to those found in other oceans and seas. This type of biomarker seems promising, and a validation on extended scale is possible, due to its simplicity. A very good example of combining simplicity with ecological realism (complexity).

2.5. PREDICTIVE TOOLS IN ENVIRONMENTAL TOXICOLOGY (SARS)

The need for prevention gives rise to a need for prediction. In Section 1, some predictive tools relating physical and chemical properties to the environmental fate of organic contaminants were presented. There are also predictive tools for evaluating and measuring toxicity.

2.5.1. SARs and Potential Carcinogenicity

Structure-activity relationship (SAR) analysis is a critical tool in evaluating the potential environmental and human health hazard of new substances, right from the initial research and development phase. This approach is currently used in drug and pesticide design, before conducting toxicological experiments, to select those substances considered most appropriate in relation to the expected type of toxic action. In the case of carcinogenicity, it is important to know which are the structural properties that make a chemical a carcinogen.

In carcinogenesis, three different mechanisms of action are possible:

- Direct damage to DNA, by covalent binding (DNA-*reactive carcinogens*)
- Indirect damage to DNA, after substance activation by chemical or biochemical processes
- Neither DNA direct nor indirect damage; cancer is produced as a result of unclear extrachromosomal mechanisms (hormonal imbalance, inhibition of intercellular communication, etc.)

In the first two cases, the chemicals are called *genotoxic carcinogens*, and are generally positive to mutagenic *in vitro* tests; those chemicals causing neither direct nor indirect damage to DNA are called *epigenetic carcinogens* and are currently negative in tests for genotoxicity.

There are some important physicochemical properties for determining the bioavailability of the substance on the target cell (Woo and Arcos, 1989). These are applied to SAR analysis for genotoxic carcinogens:

- Molecular weight: values higher than 1000 a.m.u. (molar mass > 1,000 g/mol), although with some exceptions, indicate a low probability of reaching the target
- Vapor pressure: a high vapor pressure favors volatilization and the possibility of direct inhalation
- Solubility in water: when high means rapid excretion potential and may greatly reduce carcinogenicity
- Chemical stability: if highly unstable, the carcinogenicity potential is reduced
- Molecular geometry: influences the ability to reach and react with the molecular target and chances of being metabolically activated or detoxified

In the case of epigenetic substances, due to the great differences in possible mechanisms of action, these need to be known in advance.

Carbamate pesticides, general formula:

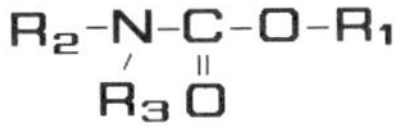

$$R_2-\underset{\underset{R_3}{/}}{N}-\underset{\underset{O}{\|}}{C}-O-R_1$$

carcinogenic when:

– R_2, R_3 = H and R_1 = vinyl, ethyl or methyl

– R_2, R_3 = H or alkyl and R_1 = $-\underset{\text{aryl}}{\overset{\text{aryl}}{C}}-C\equiv C-H$

– R_2, R_3 = CH_3 and $O-R_1$ replaced by $-Cl$

mutagenic when:

– R_2 or R_3 is acyloxy

Figure 2.7. Structural features affecting carcinogenicity and/or mutagenicity of carbamate pesticides. (From Woo and Arcos, 1989. With permission.)

In Figure 2.7, an example of the application of the SAR approach to the evaluation of carcinogenicity of carbamate pesticides is shown. The data in the Figure 2.7 suggest that, in carbamate, there are three potential electrophilic sites: the alkyl group, the carbamoyl group, and the activated amino end. Another way for carbamate pesticides to originate probable carcinogenic compounds is related to the possibility of formation of N-nitroso-carbamate, by reaction with nitrosating agents (e.g., the nitrite in saliva) in conditions favorizing N-nitrosation reactions (e.g., the acidic environment of the stomach). This example shows the possibility of predicting, in a qualitative way, mutagenic or carcinogenic effects of new substances by means of the SARs.

2.5.2. Nonspecific Toxicity and QSARs

Meyer (1899) and Overton (1901) found an increase of the narcotic effect by increasing the chain length of a group of chemicals from a homologous series. Overton explained this effect as due to the lipid/water partition: the higher the partition, the greater the effect. These findings were based on the idea that a quantitative relationship could exist between molar or molecular properties and toxicological properties. These relationships are called *quantitative structure-activity relationships*, QSARs (Calamari and Vighi, 1990) and differ from SARs because of their aim to be quantitative.

After the initial findings, the use of formal structure-activity relationships started with the pioneering work of Hammett (1940), followed by the works of Taft (1956) and Hansch and Leo (1979). The main application was in drug design and pesticide research and, in the 1970s, they began to be applied to aquatic toxicology.

In the QSAR approach, it is important to distinguish between *reactive* (or *specific*) and *nonreactive* (or *nonspecific*) *toxicity*. The former is associated with a particular chemical reaction (e.g., the inhibition of a metabolic pathway); the latter is not related to any particular interaction and is only dependent on the quantity of toxic substance acting upon the cell (Blum and Speece, 1990). QSARs for reactive toxicology are mainly applied in pharmacology, while QSARs for nonspecific toxicology are of increasing concern for environmental sciences.

As reported by Albert (1985), Ferguson extended the theory by Meyer and Overton by applying thermodynamic principles to drug activity: as the cell operates in a polar medium (water), the toxic effect of a narcotic or a depressant, essentially due to nonspecific action, is related to its activity in the water phase *(Ferguson's Principle)*. This activity is the fraction of saturation, or the ratio between the concentration of the chemical and its solubility in water. In other words, an increase in water solubility will give an increase in the dose to obtain the same amount of chemical interacting with cells. This relationship can be applied to a homologous series up to a certain limit (called the *Ferguson cut-off*), where the less water soluble substances (i.e., the more hydrophobic) are not toxic, even at saturation. A significant improvement in QSAR development was reached when Hammett (1940), studying the reactivity of substituted benzene derivatives, found a relationship with the nature of the substituent:

$$\log\left(K / K_0\right) = \rho\sigma \tag{2.12}$$

where: K is a rate or equilibrium constant for a substituted aromatic compound; K_0 is a rate or equilibrium constant for the unsubstituted aromatic compound; ρ is parameter depending on type of reaction (taken as 1 for the ionization of benzoic acid); and σ is a constant depending on the type of substituent.

Equation 2.12 may be written as follows:

$$\log K = \log K_0 + \rho\sigma \tag{2.13}$$

which is a straight line with intercept log K_0 and slope ρ. The knowledge of the constant σ will give the reactivity or partition of an unknown compound.

Taft (1956) introduced a steric substituent constant, E_s, studying the effects of different substituents on the acid hydrolysis of esters of aliphatic acids. The Taft steric constant is defined by the following relationship:

$$\log\left(k / k_0\right) = \delta\, E_S \tag{2.14}$$

or

$$\log k = \log k_0 + \delta\, E_S \tag{2.15}$$

where k is the hydrolysis rate constant for a substituted compound; k_0 is the hydrolysis rate constant for the unsubstituted compound; δ a parameter depending on the system under study (for the acid hydrolysis of esters of aliphatic acids, $\delta = 1$) ; and E_s a steric constant depending on the substituent. When E_s is known, it is possible to estimate the rate of hydrolysis in a given system for an unknown compound.

Similarly, Hansch (Hansch and Leo, 1979) introduced a subsituent constant, π, as follows:

$$\log(P/P_o) = K\pi \tag{2.16}$$

where P is the organic solvent/water partition coefficient for a substituted compound; P_o is the organic solvent/water partition coefficient for the unsubstituted compound; K a parameter depending on the solvent system (K = 1 for the 1-octanol/water system); and π is a constant characteristic of the substituent. For the 1-octanol/water system, Equation 2.16 can be written:

$$\log K_{OW} \text{ subst.} = \pi + \log K_{OW} \text{ unsubs.} \tag{2.17}$$

The equations of Hammett, Taft, and Hansch have been combined by Hansch in a more complete expression relating the logarithm of the reciprocal of an effective concentration, log (1/EC), to hydrophobic, electronic and steric properties of the toxicant:

$$\log(1/EC) = K_1\pi + K_2\sigma + K_3E_S + K_4 \tag{2.18}$$

where K_1, K_2, K_3, and K_4 are parameters of the multiple linear correlation.

Further developments (Hansch and Leo, 1979) have led to the introduction of an additional element in the last equation, so as to take into account the possibility that the increase of the constant π may increase the biological activity of chemicals, but not indefinitely: an excess in hydrophobicity leads to a progressive loss of activity due to the difficulty of the chemical to reach the site of action (*Ferguson cut-off*). Thus, Equation 2.18 was completed by adding the new term π^2 to obtain a parabolic relationship:

$$\log(1/EC) = K_5\pi^2 + K_1\pi + K_2\sigma + K_3E_S + K_4 \tag{2.19}$$

All the constants (π, σ, and E_s) in Hansch's equation are *molecular properties*, which means related to the molecular structure in a homologous series of chemicals. It is possible to apply the same approach by means of analogous *molar properties*: log K_{OW} for hydrophobicity, pK_a for electronic and molar mass, *M*, for steric characteristics:

$$\log(1/EC) = K_5(\log K_{OW})^2 + K_1 \log K_{OW} + K_2\, pK_a + K_3M + K_4 \tag{2.20}$$

Depending on the different relative weights the various constants may show in different groups of chemicals, Equation 2.20 may be reduced to a more simple expression.

As an example, Figure 2.8 shows the correlation found by Blum and Speece (1990) between predicted and observed IC50 (*inhibition of the growth by 50%*) for 53 different chemicals in aerobic heterotroph bacteria based on the simple correlation: $\log IC_{50} = 5.12 - 0.76 \log K_{OW}$ ($n = 53$; adjusted $r^2 = 0.82$).

Figure 2.9 shows an example of the application of the QSARs to develop a predictive correlation for the EC50-24h (immobilization of 50% of the treated animals) in *Daphnia magna* produced by exposure to a series of organotin compounds.

2.5.3. Interspecies Correlations

Species with a rapid lifecycle, small dimensions, and clear toxicity endpoints represent the optimum for toxicological testing. Typical organisms fitting these requirements are protozoa, small crustaceans (e.g., Daphnia), microalgae, and bacteria. In several studies, the toxicity findings from different species have been applied in trials to discover the *most sensitive species*. Empirical equations correlating the toxicity between two different species can be developed with available data. An example is that in Figure 2.10, showing the relation between the LC_{50} in fathead minnows, *Pimephales promelas*, and IC_{50} (in this case meaning *inhibition of 50% of the light emission rate*) of the bioluminescent marine bacterium *Photobacterium phosphoreum* (Microtox test, by Beckman Instruments, Inc.). When available, a correlation like that in Figure 2.10 may be applied to calculate the toxicity of an unknown substance with an error comparable with that of direct measurements (Vighi, 1989).

2.6. THE NEED FOR THE EVALUATION OF RESPONSES AT THE ECOSYSTEM LEVEL

The assessment of toxicity at the ecosystem level is the main objective of environmental toxicology. The traditional approach consists of the following steps:

- Measuring chronic toxicity in the laboratory in groups of organisms representative of a few species
- Extrapolating results to other species
- Indicating environmental quality criteria able to preserve the structure and functions of the target ecosystem

To attain the objective, *arbitrary security factors*, or *safety factors*, based on the experience of toxicologists, were applied to each step. For instance, a factor of 1/10 is applied to the NOEL for a test fish to consider the different sensitivity

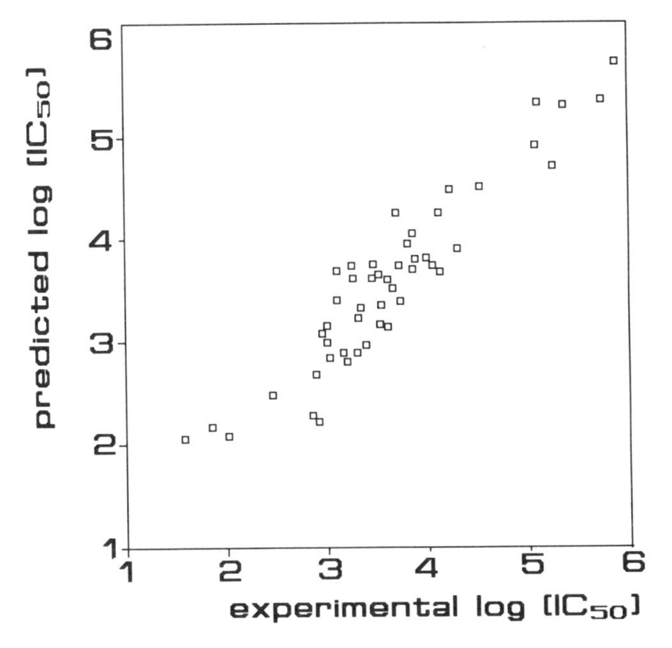

Figure 2.8. A simple QSAR for aerobic heterotroph bacteria. (From Blum and Speece, 1990. With permission.)

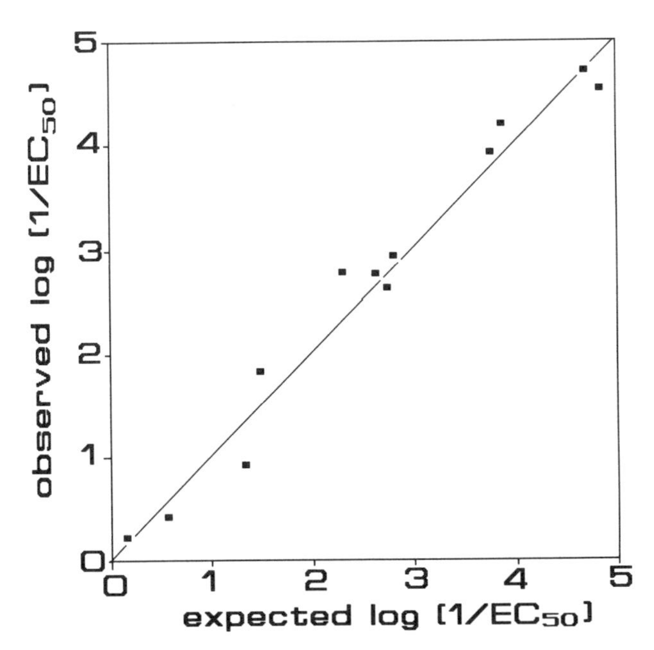

Figure 2.9. Relationship between observed and calculated reciprocal EC_{50}-24h (L/mg) in *Daphnia magna* exposed to a series of organotin compounds. QSAR equation: $\log (1/EC_{50}) = 0.412 \log K_{OW} + 0.523\ pK_a + 0.099$. (Modified from Vighi and Calamari, 1985.)

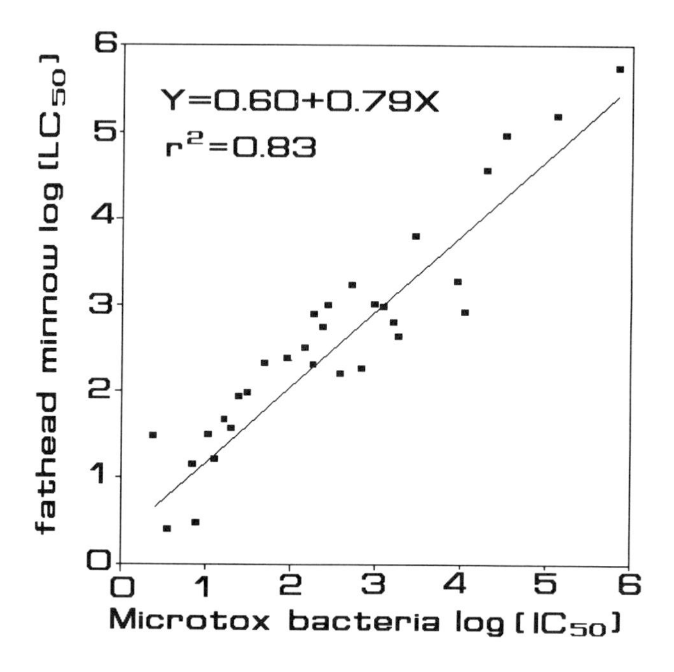

Figure 2.10. Correlation between acute toxicity (LC_{50}) in fathead minnows and Microtox bacteria. (From Blum and Speece, 1990. With permission.)

of different species, and 1/10 again to produce the water quality criteria. In general the safety factors range from 1/10 to 1/1000 (Okkerman et al., 1991).

This approach works quite well, although its limitations are obvious. However, to measure the actual impact on natural systems, the best way should be to test chemicals in the field. In the field, the system may be perfect, but the exposure, particularly if low, is very difficult to control. Ecological realism and simplicity of performance are the key parameters for ecotoxicological testing. A schematic representation of the ideal toxicological test and of the relationship between simplicity and realism is given in Figure 2.11.

In the previous chapters, simplicity was prevailing. In the next, an increase in ecological realism is presented together with its inherent complexity.

2.6.1. Field Studies (Ecosystem Monitoring)

If the 1960s was the decade of the perception of environmental pollution, the 1980s has been the decade of perception of the existence of global pollution (e.g., acid depositions, ozone layer depletion, greenhouse effect, and possible global climate changes). Consequently, the need for ecosystem monitoring has recently received great impulse, expanding the interests of toxicology, previously limited to a selected group of particularly exposed people, animal, or plants, to broader implications. This requires the development of methods to investigate the possibilities of impairment of structural and functional properties of ecosystems.

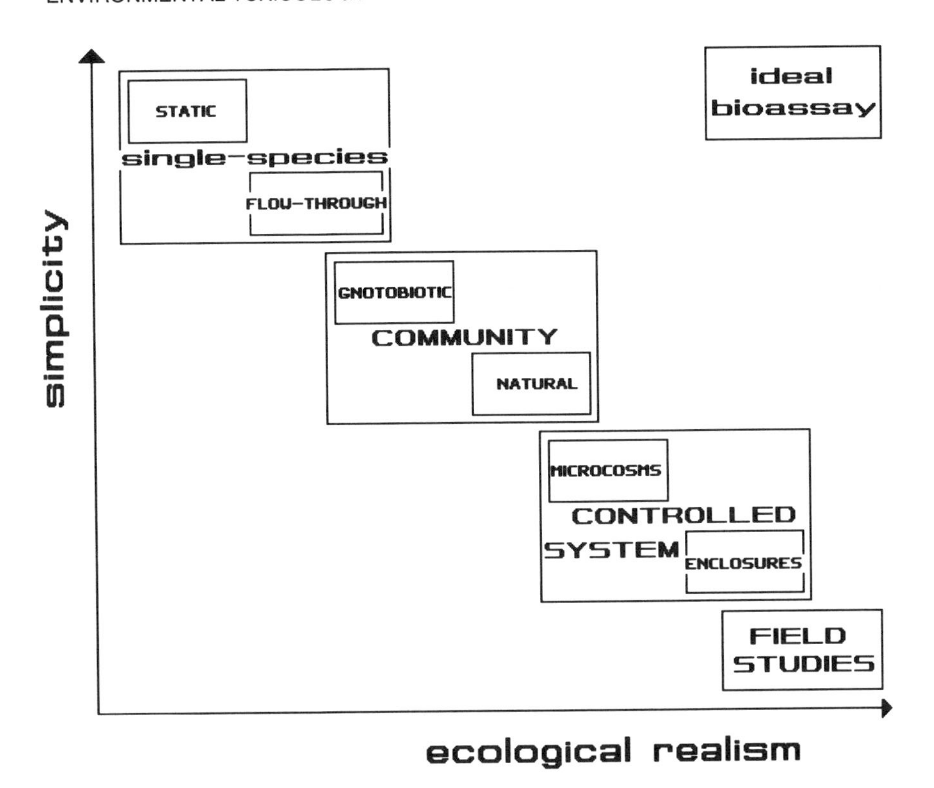

Figure 2.11. Relationship between simplicity and ecological realism in bioassays. (Modified from Blank et al., 1978; Calamari et al., 1985.)

The main difficulties in assessing potential effects on ecosystems due to a perturbation arise from the fact that these effects may occur at different levels of organization, on different space and time scales, and may interact with each other. The analysis of only one level of organization for a limited time may lead to wrong evaluations.

The ecosystem is more than the sum of its parts: it contains compensatory feedback mechanisms counteracting natural or induced perturbations (*homeostasis*). In terms of functionality, the ecosystem should be more resistant to perturbations than its most sensitive individuals and populations, since the role of one species may be played by another. However, the buffering mechanisms of ecosystems are not unlimited, and problems may arise "suddenly", after a long-term (on human temporal scale) no-effect exposure.

Effects such as the disruption of food resources, modification of habitat, and changes in competitive interactions may be dramatic for the ecosystem integrity.

Levin et al. (1984) discussed the concept of *assimilative capacity* used by Westman (1972) to define the possibility of water bodies in recovering pristine conditions after the input of a contaminant. This term, introduced by engineers to determine the ability of a water body to assimilate oxygen-demanding materials, "does not consider whether the biological integrity of the system is retained".

They correctly suggest using broader concepts, such as *resiliency*, to define the ecosystem's ability to resist structural and functional changes.

The main objective for environmental toxicology and ecotoxicology is to define the limit of resiliency beyond which significant or irreversible effects may occur.

According to Levin et al. (1984), "it is important to recognize that even small perturbations may result in changes in ecosystem structure and function, including loss of species and other components. The identification of such *catastrophic* points of system breakdown represents one of the great challenges of ecotoxicology".

Possible endpoints for these studies are measurable modifications of structural or functional ecosystem properties. Structural properties can be divided into taxonomic (e.g., population density, species composition, individual and species distribution) and nontaxonomic (conditions and resources inventory, physico-chemical characteristics). Functional properties may also be divided into taxonomic (colonization rate, predator-prey relationships) and nontaxonomic (productivity, rates of transformation processes). These possible endpoints may be measured either in the field or in field enclosures, obtained by placing artificial barriers in nature to separate a part of an ecosystem (such as a water column enclosed within plastic walls). In this way, the chemical perturbation (e.g., the concentration of a toxicant in water) may be better controlled.

Davies and Gamble (1979) enclosed within plastic up to 1300 m^3 of seawater, containing a complete natural community with plant, herbivore, and small carnivore representatives, able to sustain themselves for several weeks. Then a toxicant was added and *results indicated that the major observed effects* (e.g., modification in zooplankton density) *were due to the enclosing operation*. These and similar results indicate the complexity of problems when studying effects at the ecosystem level.

The value and limitations of some recent approaches for aquatic systems have been discussed by Ladner et al. (1989). To combine feasibility with ecological realism, these authors put forward some suggestions on the type of strategy to be selected according to the particular aim for which the system is planned:

- *Pelagic bags*: these are open on top, at the water surface, and have no flow-through system; the experiments may be carried out for a relatively short time (a few months), due to the development of *Aufwuchs* communities* on the walls, alteration of phytoplankton sinking rate, and nutrient cycling.
- *Littoral zone bentos enclosures*: these have a flow-through apparatus and may be observed over a long period of time.
- *Profundal zone enclosures*: are applied to measure some basic ecosystem functions (e.g., nutrient regeneration, respiration rate); have flow-through apparatus and may be observed over a long period of time.

* The aufwuchs is the ensemble of organisms growing attached to, or clinging upon, free surfaces of submerged objects such as stems, roots and leaves of living plants, rock, wood and animals (cfr. Newman and McIntosh, 1989).

- *Limnocorrals*: enclosures combining benthic and pelagic communities, with relatively rapid water turnover, suitable for long-term studies.
- *Experimental ponds*: natural small lakes or ponds; the only limitation is that these systems may be particular and of limited predictive potential.

Another, more classic but still little exploited approach is the use of artificial substrata in studying the rate of colonization and the species composition after reaching a sort of steady state. In aquatic systems, this can be applied either with animals or algae, or both. The test is carried out in the field; if well calibrated (i.e., along a concentration gradient from a point source), this could be an excellent tool for evaluating community responses to nutrients and pollutant stress.

2.6.2. Laboratory Models

When field enclosures are transplanted to the laboratory, as in the approach of the so-called "microcosms", the pollutant doses are more easily controlled. Microcosms may be simplified ecosystems which include some of the factors missing in single-species testing; another type of microcosm is that obtained by transferring to the laboratory a piece of a natural ecosystem. A review on terrestrial microcosm application was done by Gillet (1980).

Microcosms, like the field enclosures, have some limitations: the type and number of ecological interactions cannot include all the interactions of natural systems, essentially because of space and time scale reduction, generating a sort of *wall effect* and impeding external exchanges. Microcosms can reach good *precision*, in the sense that the replicability of results is good; as far as *accuracy* is concerned, or the ability to characterize real-world effects in a reliable way, microcosms are less satisfactory.

These approaches are more realistic than the rat or the *Daphnia* model. However, they are not free from limitations, mainly due to the reduction in scale and also to the short time span of observations. The reduction in scale, in particular, may alter the fate of contaminants and, in some cases, may induce perturbations higher than those caused by the toxicant, as in the case of open field enclosures previously mentioned. Microcosms, or model ecosystems, appear to have limited applicability due to their "*excess of simplification to be realistic and excess of complication to be interpretable*" (Moriarty, 1983).

More controlled systems are those where the number of species is dramatically reduced. *These may be applied more effectively for the study of the transport and fate of contaminants (distribution, degradation, bioconcentration, and biomagnification) than for evaluating ecological effects*. For these aspects, microcosms are much more suitable and reliable, at least in the production of relative data (i.e., comparisons of alternative chemicals). Metcalf and co-workers (1971) introduced a simplified model laboratory ecosystem which was subsequently applied to simulate the fate of more than 100 chemicals (see Figure 1.1, Section 1).

An interesting approach proposed by Ladner (1989) is that of tests with natural associations of periphyton and phytoplankton: small samples, derived from natural communities, are transferred into the laboratory where the photosynthesis rate, as toxicological endpoint, is measured in acute toxicity tests. In this way, a functional response of a natural community may be obtained in relation to known exposure levels. The limitation of this approach is the same as that of acute toxicity tests: neither direct slow effects nor indirect effects due to ecological interactions can be predicted.

One the main problems in the evaluation of the effects at the ecosystem level, making the difference with the more traditional toxicological evaluation for humans, is the temporal scale (Burger and Gochfeld, 1992). For man, the worst case is full-time exposure over the life span, which is well known and limited to 70 to 80 years. For an ecosystem, the lifetime is unknown and recovery from a damage status may be complete or incomplete and may require a very long time (on human scale). Besides, for humans, the death of even a small portion of a population is important; for an ecosystem, the disappearance of some individuals and even of some species may not be significant to its functionality.

2.6.3. Statistical Methods: The Hazardous Concentration Approach

"Single species toxicity tests are now, and probably will continue to be the backbone of our efforts to determine the probability of harm to more complex systems." This is the opinion of John Cairns, Jr. (1986), who goes on to say, *"I am deeply disturbed, however, when people use simple, short-term laboratory tests to infer a degree of protection to the environment that is not warranted by the evidence at hand."*

The article was titled *The Myth of the Most Sensitive Species*: the small number of species used in laboratory toxicity tests with respect to the total number of possible exposed species means that the probability of including the most sensitive one among tested species is remote. Furthermore, in general, it is unlikely that the most sensitive species for one toxicant is also the most sensitive for another.

The effects and responses of ecosystems are complicated by variations of the exposure level and routes, chemical distribution and transformation processes, and mixture effects. Taking the chemical exposure as constant, the effects will vary as a function of several biological and ecological parameters such as life stage, feeding conditions, stress factors, and species composition. In the same site, closely related species may be exposed in significantly different ways because of differences in behavior, nutritional habits, etc. To overcome these potential obstacles to accurate assessments, extrapolation procedures have been proposed to predict ecological effects starting from toxicity data on single species (Blank, 1984; U.S. EPA; 1984; Van Straalen and Denneman, 1989). This means that by taking a few NOEL values on the basis of their "representativeness" of different ecological groups, it is possible to calculate an *ecological NOEL* for the biological communities of each principal environmental compartment (i.e., water, soil).

Despite some intrinsic limitations, these approaches represent a way to evaluate the ecological effects of toxic substances. Okkerman et al. (1991) have recently reviewed the available techniques. The most interesting and feasible is that proposed by Van Straalen and Denneman (1989), which originates from a previous work by Kooijman (1987). Kooijman developed a model that allows for differences between species exposed to a single toxic substance, and also allows for uncertainty due to extrapolation based on a limited number of test species. *The objective of Kooijman's approach was to estimate a safety factor to protect the most sensitive species in a given biological community from the lethal effects of a chemical.* With this method, it is assumed that the LC_{50} of species in a community follows a continuous, symmetrical distribution on a logarithmic scale. As a result, particularly in large communities, the LC_{50} of the most sensitive species may be extremely low, thus requiring a very large safety factor. Besides, this method requires the number of species in the community (which, in principle, is not known) and introduces in that way an element of arbitrariness.

> *Van Straalen and Denneman (1989) modified Kooijman's model, changing the objective: to protect the functionality of an ecosystem, it may not be necessary to protect even the most sensitive species, because of the ecosystem's regulatory capacity and resiliency. If this is accepted, the safety factor is not sensitive to the extreme tail of the frequency distribution and is not dependent on the number of species to be protected. Instead of LC50 data, NOECs are indicated as the more appropriate parameters for ecological effect estimations.*

The selection, or the production *ad hoc*, of single-species NOEC should take into account the ecological functions, anatomical design, and exposure routes of test organisms. The method, originally proposed to protect organisms in soil systems, suggests normalizing NOEC data to a standard soil composition. When the number of NOEC data, m, is defined, an arbitrary δ_2 value has to be selected, indicating the probability of overestimating the hazardous concentration; another arbitrary parameter is δ_1, or the fraction of the species not protected; δ_2 and m are used to find the corresponding d_m value from Kooijman (1987; Table 2.3). Then, a safety factor, T, is derived allowing for differences among species:

$$\mathrm{T} = \exp\left\{\left(3s_m d_m / \pi^2\right)\ln\left[\left(1-\delta_1\right)/\delta_1\right]\right\} \tag{2.21}$$

where s_m = the standard deviation of ln(NOEC) values
d_m = the factor depending on the number of species tested, m, and on δ_2 (Table 2.3)
δ_1 = the fraction of the species not protected

The hazardous concentration to the fraction p of species not protected, HCp, is:

$$\mathrm{HCp} = \overline{\mathrm{NOEC}} / \mathrm{T} \tag{2.22}$$

where $\overline{\mathrm{NOEC}}$ is the geometric mean of NOEC values. For $\delta_1 = 0.05$, HCp becomes HC_5, *or the hazardous concentration to 5% of the ecosystem species.* Units of HCp are the same as for NOEC.

Table 2.3. Kooijman's d_m Values, Such That the Probability That the Sample Standard Deviation, s_m, Based on m Trials from the Standard Logistic Distribution Exceeds d_m is δ_2

	δ_2		
m	0.01	0.025	0.05
2	5.09	4.15	3.72
3	4.58	3.87	3.40
4	4.25	3.60	3.22
5	3.99	3.42	3.06
6	3.74	4.25	2.93
7	3.52	3.11	2.82
8	3.34	2.99	2.72
9	3.20	2.90	2.65
10	3.09	2.83	2.59
11	3.01	2.77	2.56
12	2.95	2.73	2.53
13	2.91	2.70	2.51
14	2.88	2.67	2.50
15	2.86	2.65	2.49
20	2.76	2.56	2.44
30	2.62	2.42	2.30
∞	1.814	1.814	1.814

From Kooijman, 1987. With permission.

The model may also be used inversely to calculate the percentage, q, of protected species for a given concentration, C:

$$q = 100\left\{1 - \left[1 + \exp\,(A)\right]^{-1}\right\} \tag{2.23}$$

with

$$A = \left[\pi^2\left(x_m - \ln C\right)\right] / \left(3 s_m d_m\right) \tag{2.24}$$

where x_m is the mean of ln(NOEC) values.

A sensitivity analysis of the model indicates that as the number of NOEC data increases (test species), the uncertainty decreases, thus producing higher hazardous concentration estimates. In other words, the use of a small number of NOEC (with high uncertainty) leads to HC*p* estimates lower than those obtained with more NOEC data. This, for environmental protection, is more appropriate than the contrary (as it is in the case of NOEC data; see Chapter 2.2.2.3).

A very simple BASIC program to calculate HC*p* and *q* is given in Appendix 2.

A modification of the van Straalen and Denneman (1989) approach, to reduce uncertainty (and the related tendency to underpredict the hazardous concentration value) due to the generally limited number of NOEC data used for calculating HCp, has been proposed by Aldenberg and Slob (1993). The authors introduced the HC*p*50, or the one-sided 50% confidence value of the hazardous concentration, by which the probabilities of underpredicting and overpredicting HC*p* are equal.

Statistical methods are still probably affected by the same limitations which are intrinsic to the NOEC estimations. These originate, as previously discussed, from the probabilty of everstimating NOEC by means of false negative findings. However, once the NOEC limitations are overcome, perhaps by means of some long-term effective concentration, statistical methods may represent an essential tool in estimating expected effects in biological communities. In other words, it is time to stimulate toxicological research towards approaches for proofing hazard instead of safety of chemicals, as discussed in Chapter 2.2.2.3. Even with these limitations, statistical methods constitute a substantial advancement with respect to previous approaches, essentially derived from classic toxicology, and unsuitable for multispecies toxicological evaluations.

2.6.4. The Keystone Species Approach

The concept of *keystone species* was introduced by Paine (1966, 1969) in describing the significance of the starfish *Pisaster ochraceus* in structuring the rocky intertidal communities on the Pacific Northwest of the U.S. After removing the top invertebrate predator, it was observed that significant changes occurred in the biological community (increase in population density of a few species, and reduction in species richness), resulting in major reorganization of the community at lower levels. Paine's conclusion was that the starfish had a disproportionate role in structuring the community and called the starfish a "keystone predator", which is a role similar to that of the keystone in the roman arch: if removed the arch collapses.

When a key predator is eliminated, a sequence of chain effects may be produced, with dramatic modifications of the structure of the biological community.

The ecological role of keystone predators can be seen in another example: the sea otter *Enhydra lutris* disappeared from broad areas of the Pacific Northwest coast in 1911, due to excessive hunting (Estes and Palmisano, 1974). Its main food item, the sea urchin *Strongylocentrotus polyacanthus*, increased in abundance and destroyed its main food, the macroalgae *Laminaria* spp. and *Agarum cribrosum*. As a consequence of the reduction in the benthos complexity, the entire community collapsed. The reintroduction of a small number of sea otters in the Aleutian Islands (Alaska) caused a reduction in the number of sea urchins, followed by a partial recovery of the pristine benthic community diversity (Estes et al., 1978). These examples illustrate the need to better understand the links between species and the mechanisms controlling community composition, in order to evaluate the effects of pollutants in ecosystems (Levin et al., 1984; Petersen and Petersen; 1989).

Keystone species may also be removed by pollution. This is the case, for example, of the seagrass *Posidonia*, distributed along the western and southern coasts of Australia (*P. australis*, *P. sinuosa,* and *P. angustifolia*) and along the coasts of the Mediterranean Sea (*P. oceanica*). Seagrass prairies, with their high biomasses (500–1000 g/m^2) and annual productivity (>1000 g/m^2; Hillman et al., 1989), are keystone components of coastal marine ecosystems, not only for the

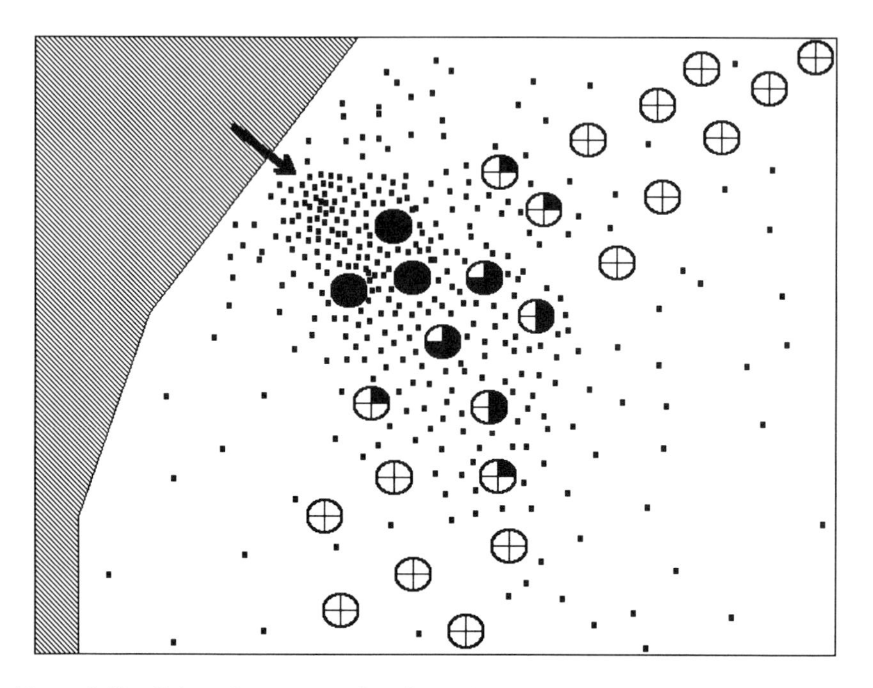

Figure 2.12. Schematic representation of a marine coastal area polluted by a point source affecting to various degrees a keystone species (e.g., the seagrass *Posidonia*).

numerous biological species directly or indirectly depending on them, but also for maintaining delicate sedimentological equilibria in the littoral (Boudouresque and Meinesz, 1982). Pollution and, to a lesser extent, the reduction of sedimentary transport by rivers due overexploitation of river beds to collect sand and gravel are the main causes of seagrass losses. Other factors have been indicated in the reduced light penetration due to an increase of suspended solids in the water column, or in an excess of epiphyte colonization on leaf and stem surfaces (Walker and McComb, 1992). It is important to point out that *seagrass degradation is often occurring in waters where the standards based on human safety are not exceeded.* These facts clearly indicate that present anthropocentric approaches are unsuitable for protecting the environment from pollution damages. On the other hand, the standardization of human lifestyle makes the world more and more homogeneous, in both technologically advanced and developing countries, with pollution problems growing in extension but reduced in variety. Urban waste waters are one of the main causes of the modification of coastal aquatic ecosystems. Water quality criteria for urban sewages, both treated and not treated, must be able to protect keystone species. To reach this objective, selected tracers could be used, such as fecal sterols and ketones (Grimalt et al., 1990), linear alkylbenzenesulfonates, and linear alkilbenzenes (Takada et al., 1992), to obtain a map of average pollution levels. This could be done in an area where pollution has been almost constant in the last 20 to 30 years and characterized by a partial disappearance of the keystone species (Figure 2.12). The quantification of pollution intensity (annual average concentration of selected tracers) and of biological and ecological damage (e.g., by measuring the reduction in population density of the keystone species) could result in a sort of response/concentration relationship, on a sufficiently adequate temporal scale (Bacci, 1993).

REFERENCES

Abram, F.S.H., The definition and measurement of fish toxicity thresholds, in *Advances in Water Pollution Research*, Vol. 3, Pearson, E.A., Ed. Water Pollution Control Federation, Washington, D.C., pp. 75-95 (1967).

Alabaster, J.S., Calamari, D., Dethlefsen, V., Konemann, H., Lloyd, R., and Solbé, J.F., Water quality criteria for European freshwater fish, *Chem. Ecol.*, 3: 165-253 (1988).

Albert, A., *Selective Toxicity, The Physico-Chemical Basis of Therapy*, 7th ed. Chapman and Hall, London (1985).

Aldenberg, T., and Slob, W., Confidence limits for hazardous concentrations based on logistically distributed NOEC toxicity data, *Ecotoxicol. Environ. Saf.*, 25: 48-63 (1993).

Ames, B.N., The detection of chemical mutagens with enteric bacteria, in *Chemical Mutagens*, Hollaender, A., Ed. Plenum Press, New York, pp. 267-281 (1971).

Anderson, E.L., Quantitative approaches in use in the United States to assess cancer risk, in *Methods of Estimating Risk of Chemical Injury: Human and Non-human Biota and Ecosystem.* SCOPE 26, SGOMSEC 2. Vouk, V.B., Butler, G.C., Hoel, D.G., and Peakall, D.B., Eds. John Wiley & Sons, Chichester, U.K., pp. 405-436 (1985).

Armitage, P., and Doll, R., Stochastic models for carcinogenesis, in *Proceedings of the 4th Berkeley Symposium on Mathematical Statistics and Probability.* University of California Press, Berkeley, CA, pp. 19–38 (1961).

Bacci E., Misura degli effetti nei sistemi acquatici, Proceedings of the meeting *Il Servizio Sanitario Nazionale ed il Controllo delle Acque Reflue*, San Vincenzo (LI, Italy), 2 April 1993. In press (1993).

Blank, H., Species dependent variation among organisms in their sensitivity to chemicals, *Ecol. Bull.,* 36: 107-119 (1984).

Blank, H., Dave, G., and Gustaffson, K., *An Annotated Literature Survey of Methods for Determination of Effects and Fate of Pollutants in Aquatic Environments*, Swedish Board for Technical Development, Solna, Sweden (1978).

Bliss, C.I., The calculation of the dose-mortality curve, *Ann. Appl. Biol.*, 22: 134-167 (1935).

Bliss, C.I., The toxicity of poisons applied jointly, *Ann. Appl. Biol.*, 26: 585-615 (1939).

Blum, D.J., and Speece, R.E., Determining chemical toxicity to aquatic species, *Environ. Sci. Technol.*, 24: 284-293 (1990).

Boudouresque, C.F., and Meinesz, A., Découverte de l'herbier de Posidonie, *Cah. Parc nation. Port-Cros,* Fr., 4: 1-79 (1982).

Burger, J., and Gochfeld, M., Temporal scales in ecological risk assessment, *Arch. Environ. Contam. Toxicol.*, 23: 484-488 (1992).

Cairns, J., Jr., The myth of the most sensitive species, *BioScience*, 36: 670-672 (1986).

Calamari, D., Bacci, E., Focardi, S., Gaggi, C., Morosini, M., and Vighi, M., Role of plant biomass in the global environmental partitioning of chlorinated hydrocarbons, *Environ. Sci. Technol.*, 25: 1489-1495 (1991).

Calamari, D., and Marchetti, R., The toxicity of mixtures of metals and surfactants to rainbow trout (*Salmo gairdneri* Rich.), *Water Res.*, 7: 1453-1464 (1973).

Calamari, D., and Vighi, M., Quantitative structure activity relationships in ecotoxicology: value and limitations, *Rev. Environ. Toxicol.*, 4: 1-112 (1990).

Calamari, D., Chiaudani, G., and Vighi, M., Methods for measuring the effects of chemicals on aquatic plants, in *Methods of Estimating Risk of Chemical Injury: Human and Non-human Biota and Ecosystem.* SCOPE 26, SGOMSEC 2, Vouk, V.B., Butler, G.C., Hoel, D.G., and Peakall, D.B., Eds. John Wiley & Sons, Chichester, UK, pp. 549-571 (1985).

Carlson, G.P., Factors modifying toxicity, in *Toxic Substances and Human Risk*, Tardiff, R.G., and Rodricks, J.V., Eds. Plenum Press, New York, pp. 47-76 (1987).

Chen, J.J., and Kodell, R.L., Quantitative risk assessment for teratological effects, J. *Stat. Assoc.*, 84: 966-971 (1989).

Cox, E.J., Naylor, C., Bradley, M.C., and Calow. P., Effect of differing maternal ration on adult fecundity and offspring size in laboratory cultures of *Daphnia magna* Straus for ecotoxicological testing, *Aquat. Toxicol.*, 24: 63-74 (1992).

Crump, K.S., An improved procedure for low-dose carcinogenic risk assessment from animal data, *J. Environ. Pathol. Toxicol.* 5: 675-684 (1981).

Crump, K.S., A new method for determining allowable daily intakes, *Fund. Appl. Toxicol.*, 4: 854-871 (1984).

Crump, K.S., and Watson, W.W., GLOBAL 79. A FORTRAN Program to extrapolate dichotomous animal carcinogenicity data to low-dose, National Institute of Environmental Health Sciences Contract No. 1-ES-2123, Research Triangle Park, N.C., (1979).

Davies, J.M., and Gamble, J.C., Experiments with large enclosed ecosystems, *Philos. Trans. Roy. Soc. London, B,* 286: 523-544 (1979).

Deichmann, W.B., Henschler, D., Holmstedt, B., and Keil, G. What is there that is not poison? A study on the Third Defense by Paracelsus, *Arch. Toxicol.*, 58: 207-213 (1986).

DeMars, R., Resistance of cultured human fibroblasts and other cells to purine and pyrimidine analogs in relation to mutagenesis detection, *Mutat. Res.*, 24: 335-364 (1974).

Doull, J., The past, present and future of toxicology, *Pharmacol. Rev.*, 36: 15S-18S (1984).

Doull, J., Introduction, in *Toxic Substances and Human Risk*, Tardiff, R.G., and Rodricks, J.V., Eds. Plenum Press, New York, pp.3-12 (1987).

Ecobichon, D.J., The *Basis of Toxicity Testing*. CRC Press, Boca Raton, FL, (1992).

Ehrenberg, L., and Osterman-Golkar, S. Alkylation of macromolecules for detecting mutagenic agents, *Teratog. Carcinog. Mutag.*, 1: 105-127 (1980).

Estes, J.A., and Palmisano, J.F., Sea otters: their role in structuring nearshore communities. *Science*, 185: 1058-1060 (1974).

Estes, J.A., Smith, N.S., and Palmisano, J.F., Sea otter predation and community organization in the western Aleutian Islands Alaska, *Ecology*, 59: 822-833 (1978).

Flamm, W.G., and Scheuplein, R.J., Use of short-term test data in risk analysis of chemical carcinogens, in *Carcinogen Risk Assessment*, Travis, C.C., Ed. Plenum Press, New York, pp. 37-48 (1988).

Finney, D.J., *Probit Analysis*, Cambridge University Press, Cambridge, U.K., (1971).

Fossi, C., and Leonzio, C., (Eds.) *Nondestructive Biomarkers in Vertebrates*, Lewis Publishers, Chelsea, MI, in press 1993.

Gillet, J.W., Terrestrial microcosm technology in assessing fate, transport and effects of toxic chemicals, in *Dynamics, Exposure and Hazard Assessment of Toxic Chemicals*, Haque, R., Ed. Ann Arbor Science Publishers, Ann Arbor, MI, pp. 231-249 (1980).

Grimalt, J.O., Fernández, P., Bayona, J.P., and Albaigés, J., Assessment of fecal sterols and ketones as indicators of urban sewage inputs to coastal waters, *Environ. Sci. Technol.*, 24: 357-363 (1990).

Hartley, H.O., and Sielken, R.L., Jr., Estimation of "safe doses" in carcinogenic experiments, *Biometrics*, 33:1-30 (1977)

Hartung, R., Dose-response relationships, in *Toxic Substances and Human Risk*, Tardiff, R.G., and Rodricks, J.V., Eds. Plenum Press, New York, pp. 29-46 (1987).

Hammett, L.P., *Physical Organic Chemistry.*, Mc Graw-Hill, New York, (1940).

Hansch, C., and Leo, A. *Substituents Constants for Correlation Analysis in Chemistry and Biology*, John Wiley & Sons, New York (1979).

Haux, C., and Förlin, L., Selected assays for health status in natural fish populations, in *Chemicals in the Aquatic Environment. Advanced Hazard Assessment*, Landner, L., Ed. Springer-Verlag, Berlin, pp. 197-215 (1989).

Hillman, K., Walker, D.I., McComb, A.J., and Larkum, A.W.D., Productivity and nutrient availability, in *Seagrasses: A Treatise on the Biology of Seagrasses with Special Reference to the Australian Region*, Larkum, A.W.D., McComb, A.J., and Sheperd, S.A., Eds., Elsevier, North Holland, pp. 635-685 (1989).

Hodson, P.V., Blunt, B.R., and Whittle, D.M., Monitoring lead exposure on fish, in *Contaminants Effects of Fisheries*, Cairns, V.W., Hodson, P.V., and Nriagu, J.O., Eds. John Wiley & Sons, New York, pp. 87-98 (1984).

Hoekstra, J.A., and van Ewijk, P.H., Alternatives for the no-observed-effect level, *Environ. Toxicol. Chem.*, 12: 187-194 (1993).

Hoel, D.G., Statistical aspects of chemical mixtures, in pp. 369-377 (1987).

Hong, W., Meier, P.G., and Deininger, R. A., Estimation of a single probit line from multiple toxicity test data, *Aquat. Toxicol.,* 12: 193-202 (1988).

Kooijman, S.A.L.M., A safety factor for LC_{50} values allowing for differences in sensitivity among species, *Wat. Res.*, 21: 269-276 (1987).

Ladner, L., Blank, H., Heyman, U., Lundgren, A., Notini, M., Rosemarin, A., and Sundelin, B., Community testing, microcosm and mesocosm experiments: ecotoxicological tools with high ecological realism, in *Chemicals in the Aquatic Environment. Advanced Hazard Assessment.* Landner, L., Ed. Springer-Verlag, Berlin, pp. 216-254 (1989).

Levin, S.A., Kimball, K.D., McDowell, W.H., and Kimball, S.F., Eds., New perspectives in ecotoxicology, *Environ. Manage.*, 8: 375-442 (1984).

Lison, L., *Statistica Applicata alla Biologia Sperimentale*, Casa Editrice Ambrosiana, Milano, Italy (1961).

Lloyd, R., The toxicity of mixtures of zinc and copper sulphates to rainbow trout (*Salmo gairdneri* Richardson), *Ann. Appl. Biol.*, 49: 535-538 (1961).

Lloyd, R., Special tests in aquatic toxicity for chemical mixtures: interactions and modification of response by variation of physicochemical conditions, in *Methods for Assessing the Effects of Mixtures of Chemicals*. SCOPE 30, SGOMSEC 3, Vouk, V.B., Butler, G.C., Upton, A.C., Parke, D.V., and Asher, S.C., Eds. John Wiley & Sons, Chichester, U.K., pp. 491-507 (1987).

Lutz, W.K., *In vivo* covalent binding of organic chemicals to DNA as a quantitative indicator in the process of chemical carcinogenesis, *Mutat. Res.*, 65: 289-356 (1979).

Mackay, D., and Paterson, S., Spatial concentration distributions, *Environ. Sci. Technol.*, 18: 207A-214A (1984).

Metcalf, R.L., Sangha, G.K., and Kapoor, I.P., Model ecosystem for the evaluation of pesticide degradability and ecological magnification, *Environ. Sci. Technol.*, 5: 709-713 (1971).

Meyer, H., Lipoidtheorie der Narkose, *Arch. Exp. Pathol. Pharmacol.*, 42: 109 (1899).

Miller, D.R., General considerations, in Butler, G.C., Ed. *Principles of Ecotoxicology*, SCOPE 12, John Wiley & Sons, New York, (1978).

Moriarty, F., *Ecotoxicology. The Study of Pollutants in Ecosystems*, Academic Press, London, (1983).

Mount, D.I., and Stephan, C.E., A method for establishing acceptable limits for fish—malathion and the butoxyethanol ester of 2,4-D, *Trans. Am. Fish Soc.,* 96: 185-193 (1967).

Müller, R., and Rajewsky, M.F., Antibodies specific for DNA components structurally modified by chemical carcinogens, *J. Cancer Res. Clin. Oncol.*, 102: 99-113 (1981).

Newman, M.C., and McIntosh, A.W., Appropriateness of aufwuchs as monitor of bioaccumulation, *Environ. Pollut.*, 60: 83-100 (1989).

Okkerman, P.C., Plassche, E.J.V.D., Sloof, W., van Leeuwen, C.V., and Canton, J.H., Ecotoxicological effects assessment: a comparison of several extrapolation procedures, *Ecotoxicol. Environ. Saf.*, 21: 182-193 (1991).

Oris, J.T., and Bailer, A.J., Statistical analysis of the *Ceriodaphnia* toxicity test: sample size determination for reproductive effects, *Environ. Toxicol. Chem.*, 12: 85-90 (1993).

Overton, E., *Studien über die Narkose*, Gustav Fisher, Jena, (1901).

Paine, R.T., Food web complexity and species diversity, *Am. Nat.*, 100: 65-76 (1966).

Paine, R.T., A note on trophic complexity and community stability, *Am. Nat.*, 103: 91-93 (1969).

Perera, F., Biological markers in risk assessment, in *Carcinogen Risk Assessment*, Travis, C.C., Ed. Plenum Press, New York, pp. 123-138 (1988).

Petersen, R.C., Jr., and Petersen, L.B.M., Ecological concepts important for the interpretation of effects of chemicals on aquatic systems, in *Chemicals in the Aquatic Environment. Advanced Hazard Assessment*, Landner, L., Ed. Springer-Verlag, Berlin, pp. 165-196 (1989).

Pfitzer, E.A., and Vouk V.B., General considerations of dose-effect and dose-response relationships, in *Handbook on the Toxicology of Metals*, Vol. 1, 2nd ed., Friberg, L., Nordberg, G.F., and Vouk, V.B., Eds., Elsevier Science, Amsterdam, pp. 149-174 (1986).

Plunkett, E.R., *Handbook of Industrial Toxicology*, Chemical Publishing, New York, (1976).

Ramaiah, N., and Chandramohan, D., Ecological and laboratory studies on the role of luminous bacteria and their luminescence in coastal pollution surveillance, *Mar. Pollut. Bull.*, 26:190-201 (1993).

Rand, G.M., and Petrocelli, S.R., Eds., *Fundamentals of Aquatic Toxicology*, Hemisphere Publishing, Washington, D.C. (1985).

Smyth, H.F., Weil, C.S., West, J.S., and Carpenter, C.P., An exploration of joint toxic action: twenty-seven industrial chemicals intubated in rats in all possible pairs, *Toxicol. Appl. Pharmacol.*, 14: 340-347 (1969).

Sprague, J.B., Factors that modify toxicity, in *Fundamentals of Aquatic Toxicology*, Rand, G.M., and Petrocelli, S.R., Eds. Hemisphere Publishing, Washington, D.C., pp. 124-163 (1985).

Sprague, J.B., and Ramsay, B.A., Lethal levels of mixed copper-zinc solutions for juvenile salmon, *J. Fish. Res. Board Can.,* 22: 425-432 (1965).

Stara, J.F., Bruins, R.J.F., Dourson, M.L., Erdreich, L.S., Hertzberg, R.C., Durkin, P.R., and Pepelko, W.E., Risk assessment is a developing science: approaches to improve evaluation of single chemicals and chemical mixtures, in *Methods for Assessing the Effects of Mixtures of Chemicals*, SCOPE 30, SGOMSEC 3, Vouk, V.B., Butler, G.C., Upton, A.C., Parke, D.V., and Asher, S.C., Eds. John Wiley & Sons, Chichester, U.K., pp. 719-743 (1987).

Stavert, D.M., Archuleta, D.C., Behr, M.J., and Lehnert, B.E., Relative acute toxicities of hydrogen fluoride, hydrogen chloride, and hydrogen bromide in nose and pseudo-mouth-breating rats, *Fund. Appl. Toxicol.*, 16: 636-655 (1991).

Stephan, C.E., Methods for calculating an LC_{50}, in *Aquatic Toxicology and Hazard Evaluation*, Mayer, F.L., and Hamelink, J.L., Eds. ASTM STP 634, American Society for Testing and Materials, Philadelphia, pp. 65-84 (1977).

Suter, G.W., II, Rosen, A.E., Linder, E., and Parkhurst, D.F., Endpoints for responses of fish to chronic toxic exposures, *Environ. Toxicol. Chem.,* 6: 793-809 (1987).

Taft, R.W., Jr., Separation of polar, steric and resonance effects in reactivity, in *Steric Effects in Organic Chemistry*, Newman, M.S., Ed. Wiley & Sons, New York, pp. 556-570 (1956).

Takada, H., Ogura, N., and Ishiwatari, R., Seasonal variations and modes of riverine input of organic pollutants to the coastal zone. 1. Flux of detergent-derived pollutants to Tokyo Bay, *Environ. Sci. Technol.*, 26: 2517-2523 (1992).

Trevan, J.W., The error of determination of toxicity, *Proc. R. Soc. Long. (Biol.)*, 101: 483-514 (1927).

Trevors J.T., A BASIC program for estimating LD50 values using the IBM-PC, *Bull. Environ. Contam. Toxicol.,* 37: 18-26 (1986).

UNEP/WHO, *Report of the Meeting of a Government Expert Group on Health Related Monitoring*. CEP/77.6, World Health Organization, Geneva, (1977).

U.S. EPA, *Estimating "Concern Levels" for Concentrations of Chemical Substances in the Environment*, Environmental Effects Branch, Health and Environmental Review Division, Environmental Protection Agency, Washington, D.C. (1984).

van Straalen, N.M., and, Denneman, C.A.J., Ecotoxicological evaluation of soil quality criteria, *Ecotoxicol. Environ. Saf.*, 18: 241-251 (1989).

Vighi, M., *Ecotossicologia*, Edizioni Giuridico Scientifiche, Milano, Italy (1989).

Vighi, M., and Calamari, D., QSARs for organotin compounds on *Daphnia magna, Chemosphere*, 14: 1925-1932 (1985).

Wakabayashi, K., Yahagi, T., Nagao, M., and Sugimura, T., Comutagenic effect of norharman with aminopyridine derivatives, *Mutat. Res.*, 105: 205-210 (1982).

Walker D.I., and McComb A.J., Seagrass degradation in Australian coastal waters, *Mar. Pollut. Bull.*, 25: 191-195 (1992).

Weber, C.I., Peltier, W.H., Norberg-King, T.J., Horning, W.B., Kessler, F.A., Menkedick, J.R., Neiheisel, T.W., Lewis, P.A., Klemm, D., Pickering, Q.H., Robinson, E.L., Lazorchak, J.M., Wymer, L.J.:, and Freyberg, R.W., *Short-term methods for estimating the chronic toxicity of effluents and receiving waters to freshwater organisms*, 2nd ed., EPA 600/4-89-001A. U.S. Environmental Protection Agency, Cincinnati, OH, (1989).

Westman, W.E., Some basic issues in water pollution control legislation, *Am. Sci.*, 60: 767-773 (1972).

WHO, *Principles and Methods for Evaluating the Toxicity of Chemicals*, World Health Organization, Geneva, (1978).

Williams, D.A., A comparison of several dose levels with a zero dose control, *Biometrics*, 28: 519-531 (1972).

Woo, Y., and Arcos, J. C. Role of structure-activity relationship analysis in evaluation of pesticides for potential carcinogenicity, in *Carcinogenicity and Pesticides. Principles, Issues, and Relationships*, Ragsdale, N.N., and Menzer. R.E., Eds. American Chemical Society, Washington, D.C., pp. 175-200 (1989).

Give us the tools, and we will finish the job.

Winston S. Churchill

SECTION 3
Hazard Evaluation and Risk Assessment

3.1. INTRODUCTION

Hazard evaluation and risk assessment for contaminants and pollutants in complex environmental systems are becoming essential tools in environmental management. These are the ultimate objectives of ecotoxicology and, to be accomplished, they need the combination of exposure data and toxicological information. More precisely, the procedure of hazard and risk analysis implies four steps (Anderson, 1988):

- Hazard identification
- Exposure assessment
- Dose-response modeling
- Risk characterization

The first step is currently carried out by means of all available data on potential or actual adverse effects produced by the substance or the mixture. In this phase, it is important to know the nature of the toxic action of the contaminant (i.e., mutagenic, carcinogenic, or unspecific toxicity) in order to select appropriate dose-response information. In the case that only threshold effects are likely, the approach of NOEL (no-observed-effect level) or ADI (acceptable daily intake) can be applied to long-term exposure to quantify the hazard by means of margin of safety (MOS) calculations; for short-term exposure, the MOS can be obtained by comparing expected or measured levels with *acute median effective concentrations*, EC_{50}, such as the *median lethal dose*, LD_{50}. When nonthreshold effects are expected, such as the induction of cancer, dose-response relationships are applied to calculate the incidence of new cancer cases (particularly in humans) associated with a given level of exposure.

Table 3.1. Definition of Terms Currently Applied in *Hazard* and *Risk* Assessment

1. *Exposure assessment*: estimation of the quantity of a chemical contaminant reaching target organisms resulting from release, transport, and fate of a chemical in the environment. Example of endpoints: environmental concentrations, doses.
2. *Effects assessment*: identification and quantification of potential adverse effects of chemicals on individuals, populations, or ecosystems by means of laboratory testing or field observation. Example of endpoints: dose causing death, reproductive failure, reduction of species diversity.
3. *Hazard assessment*: integration of points 1 and 2 to estimate the nature and magnitude of the adverse effects resulting from the release of a chemical into the environment. Examples of endpoints: comparison of environmental concentrations, measured or predicted, with NOEL; calculation of a margin of safety, MOS.
4. *Risk assessment*: quantitative estimation of the probability of clearly defined environmental effects occurring, or expected, as a result of the exposure to a chemical contaminant. Examples of endpoints: estimation of the probability of additional cancer incidence in a population.

Modified from OECD, 1989.

The assessment of environmental fate of chemicals plays an essential role in pointing out expected trajectories of the substance in natural systems and, consequently, in indicating potential direct and indirect targets.

3.1.1. Hazard and Risk: Definitions

Hazard is not synonymous with risk: the former can be defined as the presence of a potential source of danger (i.e., a toxic substance), while risk is a more precise and quantitative concept meaning the probability that a deleterious effect will occur (White and Burton, 1980). Hazard and risk assessment imply a procedure which includes the exposure and effect assessment. In Table 3.1, current definitions of these terms are shown.

The main challenge of ecotoxicology is to provide a predictive approach for hazard and risk evaluation to ecosystems. This is a very difficult task, that is only partly feasible with currently available instruments.

The procedure for environmental hazard assessment and risk evaluation by means of a predictive approach is schematically indicated in Figure 3.1.

The level and the type of exposure, together with the nature of the toxic substance, are the keypoints in hazard assessment and risk evaluation. The approach in Figure 3.1 can be applied in predictive risk assessment: starting from physical and chemical properties of a new chemical, it is possible to predict at least the general trends of its environmental distribution and fate. Consequently, an identification of the main biological targets is feasible. As far as the toxicity measurement is concerned, this is not site- or time-specific: the same measurement, if appropriate, can be applied to a great number of case studies.

3.2. ENVIRONMENTAL FATE ASSESSMENT

Knowledge of the environmental fate of contaminants is essential in correctly evaluating the sources of contamination and may help in avoiding inadvertent

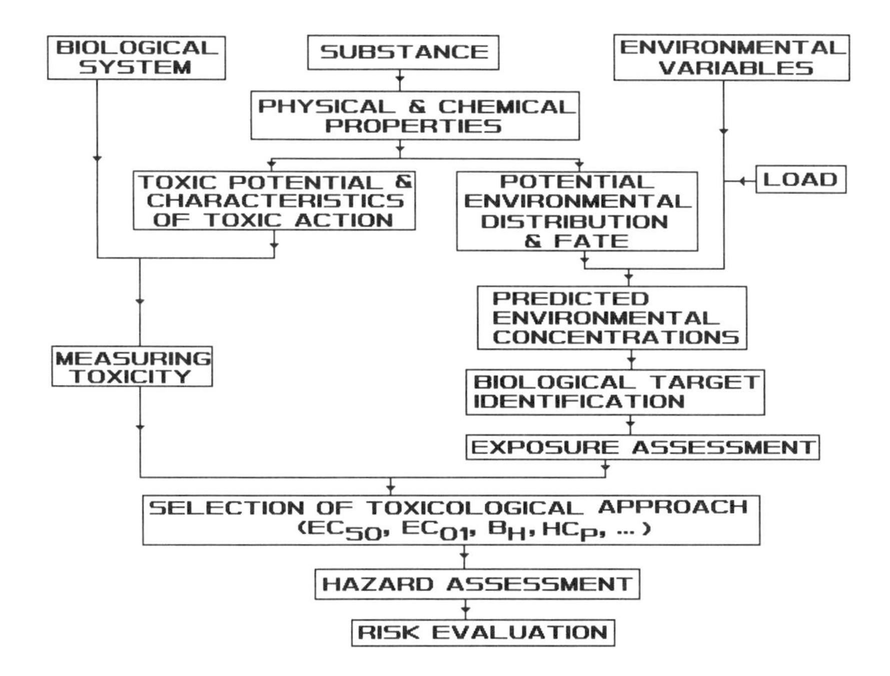

Figure 3.1. The procedure for environmental hazard assessment and risk evaluation by a predictive approach.

exposure pathways. From the other side, knowledge of the habits of the considered biological species or of the community under consideration are essential in carrying out an exposure evaluation.

3.2.1. Integrating Data from Simple Physicomathematical Models, Field Studies, and Laboratory Models

Some of the information required is often already available from existing literature or from previous studies. It is important that people involved in the "evaluation of exposure" have a direct knowledge of the site were this is to be applied. Even in the predictive approach, when contamination is not yet occurring, the site is available. Sometimes, direct contact with the site by experienced people may greatly help the work.

Another important source of information may be the use of laboratory models. Laboratory models should not be used to transfer a piece of the environment into the laboratory, since the reduction of the scale will only serve to increase the complexity of something that is already complex enough. On the contrary, laboratory experiments should be applied to measure reaction or volatilization rates, uptake-release kinetics, to calculate a bioconcentration factor or an uptake rate. In other words, laboratory models should be used as instruments for the analysis of complex processes, to say *yes* or *no*, or *how* and *why*, or to estimate *how much*. A laboratory model should be assembled on the basis of indications from a physicomathematical model, possibly extremely simple such as in the approach of the evaluative models discussed in Section 1.

Field data are needed to calibrate results from simple models (both laboratory and physicomathematical) and to obtain site-specific information. However, the analytical effort may be reduced and the interpretative potential of results enhanced if field measurements have been planned by means of a conceptual model. Even when there is a discrepancy between predicted and actual measurements, it is likely that the *relative behavior* of different chemicals in the same environment will not change. If so, this could be applied in producing criteria for chemical selection to minimize environmental contamination. This information, combined with toxicological data, may provide ecotoxicological rankings indicating *which substance is worse and which is better.*

When contamination monitoring is integrated within the framework discussed above, field data may be interpreted and applied to predict similar contamination cases in other sites or, in the same site, at different times.

3.2.2. Selection of Targets

The assessment of the environmental pathways of a chemical will indicate potential targets. For example, a pesticide with a high bioconcentration factor applied to a soil and then, by runoff, released into a lake indicates some direct targets (aquatic organisms, fish) and some indirect targets: fish-eaters (humans and non-humans). After the targets have been selected, the exposure assessment can be made. In the case of pesticide application for agricultural practices, the treated soil is not considered of primary interest for ecotoxicology: it is already accepted that an active ingredient applied to a soil could kill the target species as well as some other organisms. The environmental contamination problems originate where the treated soil ends: groundwater, surface water bodies, air. These goods are not included in the private property of the owner of the treated field. A further potential complication arises from the possibility of wild animals temporarily reaching the treated field.

3.2.3. Exposure Assessment

This has traditionally been done by direct measurement of levels of the contaminant in one or more environmental compartments. It may happen, particularly in the more traditional studies based on a retrospective approach, that certain limitations will heavily influence the quality of results, such as:

1. The "discovery" of high levels in the proximity of a contamination source
2. The scarce attention to contaminant sources, contaminant pathways, and loads
3. The lack of attention to the intrinsic properties (i.e., physicochemical) of the contaminant
4. The predominance of grid-based approaches (both temporal and spatial; e.g., a sample per kilometer or per week)

5. The risk that the investigator may have selected the case to be studied based more on his technical facilities and knowledge than on an objective priority scale
6. A high sectorialization of studies by chemical (e.g., trace metals vs organics) or environmental classes (e.g., marine vs terrestrial)
7. The use of nonindicator materials due to some homeostatic mechanism in selected organisms, or to environmental partition phenomena
8. The use of good indicators, but with an unsuitable time scale (e.g., the pursuit of a substance in an environmental phase after its displacement to another, or after its transformation)
9. The "discovery" of a contamination in a site, due to long-range transportation (e.g., some organochlorine compounds, like DDT, PCBs, etc.), and assigning it to local sources
10. The consideration of a route of exposure as the main route without controlling the weight of other routes (e.g., inhalation or respiration vs food intake or root translocation)

The list could probably be longer, but it was limited to the direct experience of the author who, particularly in the first phase of his career, applied, at least once, all of the 10 "bad principles" mentioned above.

Evaluation of the exposure requires the knowledge, when possible, of the following data on the contaminant:

1. Name and physicochemical properties;
2. Load or input and characteristics of the source (point or nonpoint; continuous or not; presence of carriers)
3. Possibility of overlapping among different sources
4. Potential environmental distribution
5. Environmental reactivity
6. Estimated environmental pathways (reservoir and sinks)
7. Biological targets
8. Route of entry into the organisms
9. Well-addressed and selected field measurements
10. Simple laboratory models to understand key processes

By applying these 10 "good principles", perhaps it is possible to obtain a reliable estimation of the exposure. These indications are not original, of course, and are reported here to address students in their initial research phase. General outlines for exposure assessment have been adopted by several governmental and international agencies. In the U.S., the Environmental Protection Agency (USEPA) published a document containing the indications reported in Table 3.2.

A particular approach in exposure assessment is by means of selected chemical markers, such as that applied by Stewart and Windsor (1990) to a sewage treatment plant effluent. Sewage effluents contribute to eutrophication of aquatic systems and may transport many potentially hazardous substances. The individuation of a constituent in the sewage plume which is unique to sewage materials may be applied to measure dilution in the receiving water, marking the transport

Table 3.2. Guidelines of the U.S. EPA (1986) for Exposure Assessment

1. General estimation on chemicals
   - Identity
   - Chemical and physical properties
2. Sources
   - Production and distribution
   - Uses
   - Disposal patterns
   - Potential environmental releases
3. Environmental pathways and fate process
   - Transport and transformation
   - Identification of major pathways of exposure
   - Predicted distribution in the environment
4. Measured or estimated concentration
5. Exposed populations
   - Human population
   - Non-human
6. Integrated exposure analysis
   - Identification of exposed populations and pathways of exposure
   - Human dosimetry and biological measurements
   - Development of exposure scenarios and profiles
   - Evaluation of uncertainty

of the sewage. The fecal sterol coprostanol (*5B(H)-cholestan-3B-ol*) is produced exclusively in humans and higher mammals by a reduction of unsaturated sterols (like cholesterol), carried out by enteric bacteria. Where chlorination has rendered the use of coliform measurements unsuitable, as is the case in several sewage treatment plants, coprostanol may be used as a tracer. A further development of the coprostanol approach was that of Grimalt et al. (1990), which suggests taking into consideration the pair coprostanol/coprostanone (*5B(H)-cholestan-3-one*) as indicator of urban sewage inputs to coastal waters.

To evaluate human exposure, information on environmental distribution and transport, and on bioconcentration potential, is essential to the estimation of both professional and nonprofessional exposure.

Essential elements in exposure evaluations are: if continuous, the duration; if pulsed and regular, the frequency. If pulsed and irregular, it is important to see if the exposure is assimilable to a single pulse with clearance. When the exposure is continuous, chronic-effect endpoints (such as the NOEL or the EC_{01}) need to be selected in damage estimation. In the case of carcinogenic substances, the incidence of added cancer cases has to be calculated. When the exposure is short, due for example to single input with rapid dissipation rate, an endpoint such as the EC_{50} may be applied.

Combining direct measurements and modeling will help provide reliable exposure data.

3.3. HAZARD AND RISK CHARACTERIZATION

Hazard and risk characterizations concern the nature and the magnitude of the hazard or of the risk to a given biological system.

3.3.1. Quantifying the Hazard

Data on the exposure pattern and on toxic potential need to be combined in order to evaluate the margin of safety and, if possible, the probability of adverse effects (risk assessment). The term *margin of safety,* MOS (Tardiff and Rodricks, 1987), has been applied to humans and defined, for a given effect, as the ratio of an experimentally determined threshold dose divided by the dose expected or measured in the system under study, EnvD:

$$MOS = LOED / EnvD$$

The same concept can be adapted to nonhumans for concentrations in the medium:

$$MOS = LOEC/EnvC$$

As an alternative, for humans, the acceptable daily intake, ADI, can be applied. This has been defined by the FAO/WHO Expert Committee on Food Additives (1974) as *"the amount of food additive that can be taken daily in the diet, even over a lifetime, without risk"*. The ADI is calculated from NOED data on animal experiments, by applying a safety factor (1/10 to 1/2000) so as to take into account differences in sensitivity to the toxic substance among different individuals (humans) and the extrapolation from toxicity data on other species. The magnitude of the safety factor is decided on the basis of the nature of the toxic action, the quality of toxicity data, and the need to extrapolate data from one species to another. The hazard is quantified by expressing the dose as a proportion of the ADI. In as much as the ADI is already calculated with a safety factor, the quantification of hazard by MOS and ADI obviously leads to different results.

An indication of the safety to different ecosystems can be taken from the calculation of HC_p values, indicating directly the fraction of species adversely affected by a given level of contamination. According to the type of exposure, the HC_p can be calculated both with acute toxicity data (e.g., EC_{50}) and with chronic toxicity data (such as the NOEL). In the case of HC_p, the hazard can be quantified by the ratio of the concentration (e.g., in water) with the HC_p for aquatic organisms. An alternative is the inverse use of HC_p calculations to obtain the proportion of species not protected with a given concentration of toxicant.

3.3.2. Estimating the Risk

Short-term toxicological data, such as mutagenicity tests, are essential in identifying the risk and in classifying the potential of chemicals for producing damage to DNA. To quantify the risk, direct dose-response or epidemiological data is needed. When data on the species concerned (generally man) is lacking, results from other species may be applied.

For carcinogenic substances, the availability of the upper-bound slope, B, and of the dose lead to the estimation of expected incidence of added cancers in selected populations (this process is currently limited to humans, but it may be applied to other species). Epidemiological data, in the case of humans, can be used to estimate the cancer risk associated with a known exposure. When this is not available, cancer potency estimates, such as the upper-bound slope values, are obtained at high exposure levels on animal models. This approach implies two kinds of extrapolations: from high to low doses and from one species to another. In addition, high intraspecific variability has been observed in humans in relation to DNA bindings and DNA repair mechanisms (see Perera, 1988). Another source of variability may arise from the interaction among different risk factors, or the so called "cocktail" problem: in the environment, single chemicals interact with each other and synergism may occur. The classic case is that of asbestos exposure and smoking: the relative risks were 5 and 11, while the combined exposure risk was 55 (Hammond et al., 1979). For chemicals with similar mechanisms of action, however, the application of the additivity hypothesis is expected to give a relatively neutral risk estimation.

The risk 10^{-6} indicates that an individual has this probability of getting cancer; considering a population of 10 million, the same risk level means 10 more cancer cases.

As previously mentioned, these calculations probably indicate risk levels higher than the actual, due to the tendency of the upper-bound slope to overestimate expected responses. For management purposes, this risk can be compared with other similar risks: the cancer risk associated with the exposure to the chemical A is 100 times greater than the risk of substance B. In risk analysis, it is important to avoid comparisons of dissimilar risks, calculated by very different approaches (such as those for transportation safety and nuclear power plants).

For noncarcinogenic chemicals, the adoption of EC (median effective concentrations) values much lower than the current EC_{50}, [such as, an $EC_{0.001}$ (incidence = 1/100,000, risk = 10^{-5}) calculated by extrapolating to low concentrations experimental data from chronic toxicity tests] may lead to a risk quantification analogous to that for carcinogens.

3.4. THE ECOTOXICOLOGICAL APPROACH. AN EXAMPLE OF APPLICATION: *SEMIVOLATILE ORGANOCHLORINATED HYDROCARBONS (SOCS) IN THE MEDITERRANEAN SEA: SOURCES, FATE, AND HAZARD ASSESSMENT*

Here an attempt to combine some of the information discussed in this book is shown, as an example of available instruments and gaps in present knowledge. The opportunity to test an ecotoxicological approach with a real problem was given by UNEP. This U.N. agency asked for an assessment document dealing with criteria for control measures to protect the Mediterranean Sea against pollution by SOCs (Bacci, 1991).

Before entering into detail, a remark: happily, notwithstanding the effort to do all the best by applying relatively new instruments to some "old" chemicals, after only two years the cut of the discussion appears already quite oxidised: this means that Ecotoxicology is going on, fast.

Semivolatile organochlorinated hydrocarbons (SOCs) include several insecticides and biocides (e.g., DDT and its derivatives, aldrin and dieldrin, heptachlor and its epoxide, lindane and its isomers, toxaphene), and substances related to chemical and technological development such as hexachlorobenzene, polychlorinated biphenyls, dioxins, and dibenzofurans (HCB, PCBs, PCDDs, and PCDFs, respectively). The denomination "semivolatile" indicates that these substances, characterized by liquid or subcooled liquid vapor pressure values from 10^{-5} to 10^{-1} Pa (20 °C), are of low volatility. However, most of these chemicals are recalcitrant to degradation, and their long life makes volatilization phenomena significant.

Concern about SOCs began in the 1960s, when DDT residues were found almost everywhere: in soils never treated with insecticides (Cole et al., 1967); in birds and seals that never leave the Antarctic (Sladen et al., 1966); in animals and human tissues (Quinby et al., 1965; Woodwell et al., 1967); in rain (Abbott et al., 1965); and in air in remote sites of the world (Risebrough et al., 1968). The "*spurious peaks*" frequently found in biological samples for DDT residue analysis were identified by Jensen (Anonymous, 1966) as components of the PCB mixtures, mainly used in electrical capacitors and transformers (considered as "closed" systems).

In the second half of the 1960s, another widespread contaminant, hexacholorobenzene (HCB), was identified. This compound essentially originated from nonagricultural sources (Heinisch, 1985). During the 1970s and 1980s, an extensive research effort revealed the significance of environmental contamination by other "DDT-like" compounds, originating from intentional applications (i.e., agricultural use, vector control operations) and unintentional dispersion (PCBs) or unintentional production (HCB). Typical "unwanted" chemicals include PCDDs and HCB from combustion processes.

All of these facts prompted the Mediterranean countries to move against Mediterranean Sea contamination in cooperation with certain U.N. Agencies (FAO, WHO, and IAEA) and under the coordination of UNEP. The Governing Council of UNEP defined the Mediterranean as a priority area, and launched the Mediterranean Action Plan (MAP; UNEP, 1982). Under MAP, extensive monitoring data on DDT and related compounds (DDTs), PCBs, and other chlorinated hydrocarbons in marine organisms were collected and published (UNEP/FAO, 1986; UNEP, 1986), leading to a first assessment document on the state of pollution of the Mediterranean Sea by these compounds (UNEP/FAO, 1990).

Starting from available information, it is possible to:

- Identify the main uses and potential contamination sources of PCBs, HCB, HCH isomers, and DDTs in the Mediterranean system
- Indicate main trends in the environmental fate of these SOCs
- Summarize present knowledge on the contamination status of the Mediterranean

- Make a hazard assessment for marine organisms, fish-eating birds, aquatic mammals, and humans
- Indicate criteria for environmental recovery and restoration as a support for new control measures

3.4.1. Production and Uses

3.4.1.1. PCBs

The polychlorinated biphenyls (PCBs) are a class of compounds derived by chlorination of the biphenyl. They are characterized by excellent dielectric properties, high chemical stability, and relatively low volatility. Their industrial use began in 1929 in hydraulic fluids, plasticizers, adhesives, wax extenders, heat transfer systems, carbonless reproducing paper, and especially electrical capacitor and transformer dielectric fluids (Hutzinger et al., 1974). In the 1970s, dispersive uses were essentially reduced to so-called "closed systems"; that is, dielectric fluids for capacitors and transformers. Between 1929 and 1977, the total U.S. production of PCBs was 610,000 metric ton (Miller, 1982). In 1979, the USEPA banned their manufacture and use in the U.S. Other industrial countries were involved in PCB production: for example, Japan (production stopped in 1976), the U.S.S.R., the United Kingdom (production stopped in 1977), Germany (production stopped in 1983), and three Mediterranean countries (France, Italy, and Spain) (Geyer et al., 1984). The 1973–1980 production of PCBs by France, Italy, and Spain was of the order of 100,000 metric ton (Geyer et al., 1984). The general trend is to replace PCBs with less hazardous substances; another less official trend is to move production plants and consumption to less regulated countries (i.e., developing countries). At present, total worldwide PCB production can be estimated at 2 million metric ton (Hansen, 1987).

As far as the Mediterranean is concerned, there are no published figures on PCB production or import other than those mentioned above. However, the 100,000 metric tons already mentioned is an indication of the order of magnitude of PCB utilization in the Mediterranean area.

3.4.1.2. HCB

Commercial production of HCB began in 1933, primarily as a seed dressing for wheat to replace highly toxic mercurial fungicides. It has also been used as a wood preservative, polymer additive, in dyes, pyrotechnic products, and flame-retarding additives (Burton and Bennet, 1987). In 1978, intentional production of HCB as fungicide and wood preservative was stopped in the U.S.; these uses have been restricted in several other countries. However, the main production of HCB is essentially unwanted: HCB forms during the production of other chlorinated hydrocarbons (tetrachloroethylene and various pesticides). For instance, technical preparations of the fungicide quintozene normally contain 1 to 6% of HCB (Peattie et al., 1984). Another unwanted source of HCB is from combustion processes, mainly hazardous and municipal waste incineration: 0.8 to 4.4 t/a was the estimated release rate in the Federal Republic of Germany (Rippen and Frank, 1986); however, this is probably a minor portion of total annual HCB production. This seems to be of the order of 10,000 t/a on a worldwide basis, with about 50 to 80% of the production in Europe (Rippen and Frank, 1986).

3.4.1.3. HCHs

Hexachlorocyclohexanes (HCHs), formerly improperly called BHC (benzene hexachloride), are eight isomers, only one of which has insecticide properties: the γ-isomer, lindane. The insecticide activity of technical HCH mixtures (68–78% α-isomer, 9% β, 8% δ and only 13–15% γ-HCH; Tatsukawa et al., 1972) was discovered in 1939 and introduced into the world market immediately after the Second World War (Morrison, 1972). In more technically advanced countries, several restrictions on the use of lindane have recently been adopted and the old formulations replaced with others containing about 95% γ-HCH. Lindane, however, is still in use, especially in developing countries. The total global production of HCHs is estimated at 2 million metric tons. In Mediterranean countries, the application rate of HCH was of the order of 2000 t/a during the 1970s (UNEP/FAO, 1990). Use of these compounds has now decreased. The total application in the Mediterranean countries probably amounts to 40–50,000 metric tons HCH isomers.

3.4.1.4. DDTs

The insecticidal properties of 2,2-bis-(*p*-chlorophenyl)-1,1,1-trichloroethane, or *p,p′* -DDT, were discovered in 1939 (Metcalf, 1973). Industrial production began in 1942, reaching a maximum in the U.S. in 1963 (81,300 t; Woodwell et al., 1971); during the 1960s, about 70% of the U.S. production was exported. Total global production is of the order of 3 million tons (Woodwell et al., 1971). In E.E.C.-Mediterranean countries, the use of DDT has been strongly reduced. No information is available from non-E.E.C. countries. During the 1970s, the application rate of DDT around the Mediterranean was close to 2000 t/a (UNEP/FAO, 1990), indicating that the total application in this area is similar to that of HCHs (40–50,000 t). Compounds associated with DDT as impurities in technical formulations and as main metabolites are *p,p′*-DDE and *p,p′*-DDD.

3.4.2. Environmental Distribution and Fate

The chemicals under discussion are characterized by low or very low degradation rates under current environmental conditions. Often, the degradation products are very similar in toxic potential and partition properties to the parent compounds (e.g., the series DDT, DDE, DDD). Because of the low chemical reactivity, only partition properties will be considered in describing the main trends in SOC environmental fate.

These chemicals are generally applied in terrestrial systems where the soil is, even unintentionally, the first significant target. Thus, the mobility from soil to other environmental compartments (air, water, and, to a lesser extent, organisms) is a key element in controlling environmental redistribution. As soil turnover is generally negligible, there are only two significant ways of displacement from the areas of direct impact (e.g., agricultural fields or landfills): by water and by air.

The SOCs discussed here are nonpolar and, consequently, are not easily transported by water. The only effective way to carry these substances by water is the water runoff of soils. However, this phenomenon is discontinuous and it is relatively ineffective over long distances (e.g., 100 km or more).

The situation of transport by air movement is quite different, being characterized by an enormous turnover of the air. From soil, vapor movements of trace contaminants can be expected to be significant when the vapor pressure of the chemical is as low as 10^{-6} Pa (Bacci and Calamari, 1990). In the case of chemicals which are solid at ambient temperature (20–25°C), properties such as the water solubility and vapor pressure have to be considered referring to the liquid state (subcooled liquid, for solids). This is because the high dispersion in the environment impedes the formation of crystals (i.e., the solid state).

Once in the air, the partition between gaseous and particulate phases is regulated by the subcooled liquid vapor pressure of the chemical (Bidleman et al., 1986; Mackay et al., 1986; Foreman and Bidleman, 1987): substances with a liquid vapor pressure of 3×10^{-5} Pa (25°C) can be expected to be equally distributed between air and particulate in rural environments. In urban environments, due to a higher particulate concentration, 3×10^{-4} Pa is the vapor pressure value for equipartitioning. Higher vapor pressure values will lead to an increase of the fraction in the vapor phase. So, even *p,p'*-DDT, the vapor pressure of which is 1.5×10^{-4} Pa at 20°C (as subcooled liquid; Suntio et al., 1988) migrates slowly but continuously from soil to air (where it is essentially in the vapor phase), reaching remote sites (WHO, 1989). However, soil can delay the movement of less volatile and less polar chemicals from contaminated sites to remote areas and seas. In this retardation phenomena, plant foliage may play a significant role, due to the high affinity of foliage for SOC vapors (Bacci et al., 1990a, Calamari et al., 1991).

Air has the highest carrying potential, even for long distances: vapor movement from contaminated soils to the air is probably the main path of chemical displacement, by means of a sort of global gas chromatographic process (Risebrough, 1990; Bacci et al., 1990b).

Vapor movement, together with some dry and wet deposition, can transport significant quantities of SOCs from contaminated lands to the oceans (Kurtz, 1990).

Air-water exchange is regulated by the air/water equilibrium partition coefficient, K_{AW}. It is important to remember that the *K_{AW} of water is 1.7×10^{-5}* (20°C). As a consequence, chemicals with K_{AW} values of two or more orders of magnitude lower than water are not able to pass from water to air and, if water evaporates, their concentration in water will increase (Thomas, 1990). Chemicals with a K_{AW} > 10^{-5} are able to pass from water to air, and this is the case with all SOCs being

Table 3.3. Selected Properties of Some Organochlorine Hydrocarbons

Name	Molar mass	S_L[a] (mol/m³)	P_L[b] (Pa)	K_{AW}	log K_{OW}
PCB (60% Cl)	361[c]	4.8×10^{-5}	8.4×10^{-4}	7.1×10^{-3}	6.9
HCB	284.8	1.0×10^{-3}	1.3×10^{-1}	5.4×10^{-2}	6.0
α-HCH	298.8	8.4×10^{-2}	7.3×10^{-2}	3.6×10^{-4}	3.8
γ-HCH	298.8	1.9×10^{-1}	2.5×10^{-2}	5.3×10^{-5}	3.8
p,p'-DDT	354.5	6.5×10^{-5}	1.5×10^{-4}	9.7×10^{-4}	6.0
p,p'-DDE	318	6.3×10^{-4}	5.0×10^{-3}	3.3×10^{-3}	5.7

[a] Subcooled liquid water solubility.
[b] Subcooled liquid vapor pressure.
[c] Average from the components of the Aroclor 1260 mixture.

From Bacci et al., 1990b. With permission.

considered here (Table 3.3). This water-to-air transfer is particularly easy for HCB, with a K_{AW} of 5.4×10^{-2}. However, it can only occur when the water is already contaminated with SOCs. Only if air is more contaminated than water will the flow be from air to water. Recent measurements of the net air-to-water flux of α- and γ-HCH in the Bering and Chukchi Seas have been carried out by Hinckley and collaborators (1991): mean values were 25 and 31 ng/(m²d) from air to the sea, respectively.

The processes discussed above, in addition to input by dry deposition (air-borne particulate matter) and wet deposition (precipitation), are responsible for the reversible transfer of contaminants from air masses to water bodies.

In water, there are other phases able to reversibly capture SOCs. As these chemicals are nonpolar, they can be taken up by particulate matter (both living and nonliving) and reach the sediment compartment (essentially the organic carbon fraction). Like soil in terrestrial systems, sediments constitute the main reservoir of SOCs in aquatic systems.

The main partition property regulating the adsorption of chemicals onto suspended particles and sediments, as well as the bioconcentration of hydrophobic substances, is the 1-octanol/water equilibrium partition coefficient, K_{OW} (Karickhoff et al., 1979; Mackay, 1982): log K_{OW} values of selected SOCs, ranging from 3.8 to 6.9, are shown in Table 3.3.

To conclude this section, there is scientific evidence that SOC-contaminated soils can constitute a source of SOC, capable of reaching and contaminating distant land and water bodies by air transport. These processes are theoretically reversible; however once bathyal zone sediments are contaminated, the return is probably more difficult. Another important aspect is the time needed for the recovery of contaminated water bodies; the sediment reservoir may play the role of a secondary contamination source.

3.4.3. Sources of Contamination

Agricultural runoff, rivers, and direct discharge of industrial and municipal wastes have been estimated to contribute a total organochlorine pesticide load of about 90 t/a (UNEP/FAO, 1990). No information is available for PCBs. However, as recently shown (GESAMP, 1989), atmospheric transport of SOCs from land to ocean is the major route (78 to 99% of the total input). Riverine input of SOCs into the Mediterranean is unlikely to be the main route to marine environments.

The main land-based sources of SOCs are soils treated for agricultural practices or vector control operations, landfills, production sites, and all contaminated material and apparatus in *contact with flowing air*, located in Mediterranean and non-Mediterranean countries. Thus, for this group of contaminants, the Mediterranean Sea is not the classic semiclosed system characterized by very low water turnover (as it is for some trace metals), but an open system due to the carrying potential of air masses flowing over it and connecting it with world oceans and lands. This Mediterranean Sea is essentially the upper, mixed layer, down to 200 m depth, able to exchange contaminants with the air phase and to eliminate contaminants in deeper waters. These constitute another, less-known Mediterranean, the deep-Mediterranean, below 200 m. It is a dark and almost closed system, receiving large quantities of contaminants by deposition of particulate matter (including fecal pellets from marine organisms) and unable to rapidly release them to the air and to destroy them by photochemical reactions. This second Mediterranean is probably the weakest component of the system, where *"change comes slowly, if at all"* (Carson, 1950). Direct and indirect input of stable chemicals into the deep-sea environment may trigger irreversible destructive processes.

3.4.4 Present Status of Contamination in the Off-Shore Mediterranean

In a previous study, great effort was made to collect, organize, and interpret all available information from field measurements, mainly carried out under MED POL III and MED POL Phase II (UNEP/FAO, 1990). A further condensation of available information is attempted in the present discussion, with the aim of seeing the Mediterranean as a homogeneous system. Due to their very high variability and relatively small significance in the total system, coastal areas are not considered. In other words, the question discussed is whether the off-shore Mediterranean is significantly contaminated by SOCs.

3.4.4.1. Air

Levels of PCBs in off-shore air over the Mediterranean were found to be 0.04 to 0.30 ng/m^3, with an average of 0.15 ng/m^3 (n = 11; S.D. = 0.10; UNEP/FAO, 1990; Villeneuve, 1985). PCB concentrations ranging from 0.03 to 1.00 ng/m^3, average 0.34 ng/m^3 (S.D. = 0.27, n = 35), were measured from 1975 to 1977 in samples of "near-ocean atmosphere" (Musée Océanographique de Monaco; Villeneuve, 1985). The two average values are not greatly different, indicating that the air over the off-shore Mediterranean does not differ much from the continental air. PCB levels in the latter have been recorded at 0.96 ng/m^3 in the Netherlands (Guicherit and Schulting, 1985) and a mean level of 0.83 ng/m^3

(range 0.125 to 2.600, n = 6) in Germany, near Ulm (Wittlinger and Ballschmiter, 1987). Levels in pine needles from Corsica and Sardinia (unpublished data), and from the Italian Peninsula (Gaggi et al., 1985), are typically of the order of 50 ng/g dry weight; assuming that the needle-to-air bioconcentration factor is equal to the azalea leaf/air bioconcentration factor (Bacci et al., 1990b), the air over Corsica, Sardinia, and the Italian Peninsula should have a PCB concentration of 0.54 ng/m^3.

As far as the other SOCs are concerned, no experimental data is available for the air over the Mediterranean. However, taking HCB as a tracer, it is interesting to observe that levels of this compound in the marine atmosphere are almost homogeneous, falling in the range 0.1 to 0.2 ng/m^3, in both hemispheres (Atlas and Schauffler, 1990; Bidleman et al., 1990; Calamari et al., 1991).

In principle, due to the high mixing and transport potential of air, SOC levels over the Mediterranean should not greatly differ from levels over other seas or oceans. So, as a provisional indication of levels in air, the following values can be taken: 0.288 ng/m^3 for the sum of HCH isomers (HCHs) and 0.022 ng/m^3 for DDT isomers and related compounds (DDE, DDD, *o,p'*- and *p,p'*-isomers) which can be regarded as typical global geometric mean values (Calamari et al., 1991).

3.4.4.2. Water

PCB levels measured in open Mediterranean waters range from <0.1 to 8.5 ng/L, with higher values in the surface microlayer (UNEP/FAO, 1990). Villeneuve et al. (1981) found no significant differences along vertical water column profiles (0 to 4500 m), with a mean PCB concentration of 0.76 ng/L (n = 76, S.D. = 0.5). Most of the readings of Picer and Picer (1979) from about 50 water samples from the north Adriatic were below the detection limit of the method used (0.1 ng/L), suggesting a concentration of 0.1 to 0.2 ng/L as an acceptable estimate of present levels of PCBs in Mediterranean waters. Krämer and Ballschmiter (1988) found no variation in PCB concentration with depth in a north Atlantic water column (0 to 1220 m), and an average PCB concentration of 0.175 ng/L (n = 8, S.D. = 0.112).

Data on HCB concentrations in Mediterranean waters is scarce (UNEP/FAO, 1990). Krämer and Ballschmiter (1988) found no variation in HCB concentration with depth in a north Atlantic water column (0 to 1220 m), and an average HCB concentration of 0.005 ng/L (n=8, S.D.=0.003).

For HCHs, α- and β-isomers are the dominant forms, with average levels measured in open seawaters ranging from 0.05 to 8.9 ng/L (as sum of HCHs; UNEP/FAO, 1990). Concentrations found in waters from other areas are as follows (Kurtz and Atlas, 1990): NE Pacific, 2.9–3.9 ng/L; NW Pacific, 0.9–1.7 ng/L; Coral Sea, >0.45–2.5 ng/L; and tropical S Pacific, 0.30–0.35 ng/L. Hidaka and Tatsukawa (1983) found levels of 0.2 to 0.6 ng/L in Antarctic seawaters. From these indications, a level of 0.7 ng/L can be assumed for Mediterranean waters.

Levels of DDTs in Mediterranean waters show considerable scatter, with some very high values. However, the relatively high detection limits represent a limitation to achieve reliable results. Most of the levels measured by Picer and Picer

(1979) lay below the detection limit (0.05 ng/L). A concentration of DDTs of 0.05 ng/L seems acceptable for off-shore Mediterranean waters.

3.4.4.3. Sediments

From the previous report by UNEP/FAO (1990), it can be seen that average concentrations of PCBs in Mediterranean sediments range from 0.8 to 155 ng/g (dry weight, d.w.). A simple arithmetic mean of the values reported for the 10 areas in which the Mediterranean was subdivided (UNEP/FAO, 1990) gives an average of 45 ng/g d.w. (S.D. = 52); due to the high scattering of data, essentially caused by the presence of samples from heavily contaminated sites (harbors, direct input areas), a geometric mean value seems to be more appropriate. This corresponds to 16 ng/g. Boon et al. (1985) found PCB concentrations ranging from 380 to 4700 ng/g (0.38 to 4.70 ng/g) in North Sea sediments.

There is no data on HCB concentrations in Mediterranean marine sediments. As an indication, levels measured in North Sea sediments have been 0.002 to 0.1 ng/g (Boon et al., 1985). HCHs in Mediterranean sediments: the geometric mean of the average values reported in UNEP/FAO (1990) for the above-mentioned 10 areas indicates a concentration of 1.3 ng/g.

For DDTs, data from the NW Mediterranean and Aegean (areas II and VIII) seems to reflect heavy local contamination, with concentrations up to 1893 ng/g (UNEP/FAO, 1990). The geometric mean from the other eight areas is 4.3 ng/g; this could be taken as representative of the concentration of DDTs in off-shore Mediterranean sediments. Levels ranging from 0.05 to 2.4 ng/g have been reported for North Sea sediments (Boon et al., 1985).

3.4.4.4. Fish

Selecting the striped mullet *Mullus barbatus* as representative of all aquatic organisms, levels of PCBs in Mediterranean fish can be evaluated by the geometric mean of mean values from the 10 Mediterranean study areas (UNEP/FAO, 1990). From this data, ranging from 1.4 to 557 ng/g fresh weight (f.w.), the mean value is 68 ng/g f.w..

There is no HCB data for fish. As an indication, levels for the Norwegian lobster *Nephrops norvegicus* can be considered. These range from 0.1 to 0.6 ng/g f.w., with a mean of 0.3 ng/g (UNEP/FAO, 1990). HCH levels in the striped mullet range from 0.03 to 15 ng/g f.w. (UNEP/FAO, 1990). A value of 0.3 ng/g f.w. can be indicated as representative of levels of HCHs in fish.

Average levels of DDTs: in the striped mullet range from 19 to 175 ng/g f.w.; the range of all reported data is quite large: 3 to 400 ng/g. A typical value for DDTs can be set at 10 ng/g f.w.

3.4.5. An Evaluative Model to Interpret Results

To better understand the present situation in the Mediterranean, it could be useful to try a simple equilibrium model. Contamination sources are external and

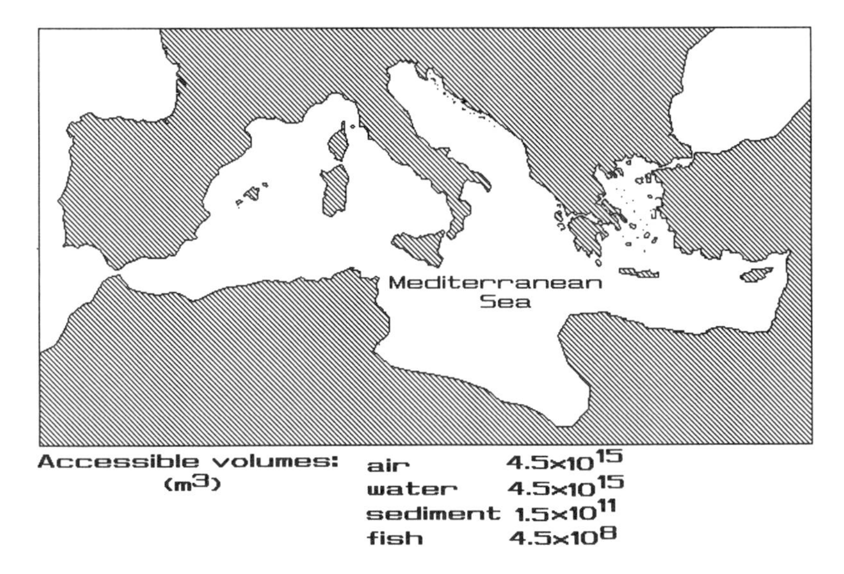

Figure 3.2. The Mediterranean: map and accessible volumes of considered compartments.

we assume that there are no significant concentration variations in the system. Only air, water, sediments, and fish (representing all aquatic organisms) are considered and the system is assumed to be closed.

Accessible volumes can be taken as follows:

- *Air: 4.5×10^{15} m^3 (surface, 3×10^{12} m^2; height, 1.5×10^3 m)*
- *Water: as for air, 4.5×10^{15} m^3 (surface, 3×10^{12} m^2; mean Mediterranean depth, 1.5×10^3 m)*
- *Sediment: Mediterranean surface (3×10^{12} m^2) for an accessible depth of 0.05 m gives 1.5×10^{11} m^3*
- *Fish: taking a biomass concentration of 0.1 g/m^3 of water and the density of fish to be 1g/cm^3, with a fish vol/vol concentration in water of 0.1 cm^3/m^3, the fish volume for the Mediterranean is 4.5×10^8 m^3 (Figure 3.2)*

Air density is taken to be 1.19 kg/m^3 and sediment density 1.5 kg/L. The organic carbon fraction (mass/mass) of sediments is considered as 0.04. Once the system is defined, along with its main properties, some basic properties of the chemicals in question are necessary to produce the simulation. These are molar mass, vapor pressure, water solubility, and K_{OW}. The data in Table 3.3 was used (for HCHs, the mean vapor pressure and solubility values between α- and γ-isomers were taken).

An equilibrium fugacity model (Level I; Mackay, 1979; Mackay and Paterson, 1982) can now be applied to simulate the distribution of PCBs, HCB, HCHs, and *p,p′*-DDT in the Mediterranean. Properties of the chemicals were selected as indicated in Table 3.3. The load of each contaminant into the system was chosen to obtain concentrations in the main environmental compartments approximating

Table 3.4. Partition, Concentration (Calculated and Measured Values) and Total Mass of SOCs in the Mediterranean System at Equilibrium (20°C)

	% Partition	Concentration	
		Calculated	Measured
		PCBs	
Air	0.066	0.74 ng/m³	0.54 ng/m³
Water	9.224	0.1 ng/L	0.1–0.2 ng/L
Sediment	90.358	20 ng/g	16 ng/g
Fish	0.352	39 ng/g	68 ng/g
		TOTAL: 5000 t	
		HCB	
Air	2.33	0.26 ng/m³	0.1–0.2 ng/m³
Water	43.64	0.0048 ng/L	0.005 ng/L
Sediment	53.82	0.120 ng/g	0.002–1 ng/g
Fish	0.21	0.233 ng/g	0.3 ng/g
		TOTAL: 50 t	
		HCHs	
Air	0.015	0.096 ng/m³	0.288 ng/m³
Water	99.210	0.66 ng/L	0.7 ng/L
Sediment	0.772	0.1 ng/g	1.3 ng/g
Fish	0.003	0.2 ng/g	0.3 ng/g
		TOTAL: 3000 t	
		DDT	
Air	0.04	0.094 ng/m³	0.022 ng/m³
Water	44.66	0.099 ng/L	0.05 ng/L
Sediment	55.08	2.4 ng/g	4.3 ng/g
Fish	0.22	4.8 ng/g	10 ng/g
		TOTAL: 1000 t	

Note: Accessible volumes (m^3): air = 4.5×10^{15}; water = 4.5×10^{15}; sediment = 1.5×10^{11}; fish = 4.5×10^{8}. Properties of chemicals from Table 3.3.

field measurement levels. In this way, it is possible to obtain a reliable indication of the total quantity of each contaminant actually present in the system, as well as its distribution. The results are summarized in Table 3.4.

The data in Table 3.4. clearly indicates that the Mediterranean system is not far from an equilibrium condition (terrestrial systems, containing contaminated soils and other sources, were excluded from the model). Thus, any significant variation in concentration in a compartment will lead to a proportional variation in *all* the others.

Available estimates of organochlorine pesticide loads (no PCBs, and essentially only HCHs and DDTs) by direct input (industrial and municipal waste waters), run-off from treated fields, and river transport are of the order of 100 t/a (UNEP, 1984). The HCHs present in Mediterranean waters and sediments are about 3000 t, and DDT is of the order of 1000 t. This data confirms the significance of air transport and indicates that if riverine input were to stop, the Mediterranean Sea would still be contaminated by long-range air transport from non-Mediterranean areas, as in the case of open oceans and remote areas.

The quantity of PCBs (5000 t) confirms these facts: the use of these chemicals is far less dispersive than the use of pesticides (lower runoff); nevertheless, very large amounts reach the Mediterranean. This is not related to a substantially different environmental load, since the production and use of PCBs is of the same order of magnitude as that of DDTs and HCHs.

3.4.6. Future Trends and Hazard Assessment to Wildlife

The partition properties of SOCs indicate that in terrestrial environments the soils act as reservoirs; although precise information on this aspect is not available, it is easy to conclude that even if production is stopped, present levels in the Mediterranean and other seas and oceans will tend to increase, or at least remain steady for a long time.

As a consequence, chronic exposure to these chemicals is unavoidable for all living organisms. To evaluate possible deleterious effects from lifetime exposure to nonmutagenic and noncarcinogenic substances, the NOEC (no-observed-effect concentration) approach may be applied. In the case of the present SOCs, which are carcinogenic in laboratory animals, the NOEC approach is not sufficient. However, in view of the complete lack of information on carcinogenicity in marine animals and plants, NOECs from selected species were used to obtain an indication of possible damage to the aquatic biological community by calculating the "hazardous concentration" 5% (HC_5). For risk analysis of humans exposed to these substances through fish consumption, current approaches for carcinogenic chemicals may be applied. As far as marine mammals and birds are concerned, simple qualitative considerations will be proposed, simply to point out the high biomagnification of these chemicals in air-breathing fish eaters.

3.4.6.1. Aquatic Organisms

3.4.6.1.1. PCBs. Commercial (and environmental) mixtures of PCBs consist of chlorinated biphenyls which differ in degree of chlorination and in position of the chlorine atoms. Both these characteristics influence the mode and intensity of toxic action (Goldstein, 1980). Thus, PCBs can be divided into two separate groups: phenobarbital-type inducers and 3-methyl-cholanthrene inducers. The former effect is related to the synthesis of cytochrome P450, the latter to the induction of cytochrome P448. Planar PCBs (those with no *ortho*-chlorine atoms) and with Cl atoms in both *meta*- and *para*-positions are cytochrome P448 inducers, like PCDDs, and in some cases highly toxic. On the other hand, the presence of *ortho*-Cl leads to induction of cytochrome P450. Therefore, in the PCB group, there are compounds of very low toxicity, together with very toxic compounds. An example of the latter is 3,4,3′,4′-tetrachlorobiphenyl, the acute toxicity of which, relative to 2,3,7,8-tetrachlorodibenzo-*p*-dioxin (2,3,7,8-TCDD, one of the most toxic substances known), is only two orders of magnitude less (single dose oral LD_{50} in the guinea pig; McConnell and McKinney, 1978).

In the environment, PCBs typically behave as a mixture, and several attempts have been made to evaluate their toxic potential by means of mixtures containing 48, 54, and 60% chlorine.

Table 3.5. No-Observed-Effect Concentrations (NOECs) of PCBs for Representative Marine Organisms and Hazardous Concentration 5% (HC_5)

Organism	NOEC (ng/L)	Ref.
Microalgae	50	Fisher and Würster, 1973
Crustacean	1000	Dalla Venezia and Fossato, 1986
Bivalve mollusc	100	Nimmo et al., 1975
Fish	100	DeFoe et al., 1978
HC_5 = 3.4 ng/L		

The NOECs for some representative marine organisms are reported in Table 3.5. These were used to calculate the HC to 5% of the species of the aquatic community (HC_5) according to the method of van Straalen and Denneman (1989). As recently discussed by van Leeuwen (1990), the toxicity data of single species can be regarded as an independent random measurement of the toxicity to the entire biological community. Hence, toxicity measurements for different representative organisms can be used to calculate the HC_p, where p is the proportion of species expected to be adversely affected by a given exposure level (i.e., concentration of the toxic substance in the water). Assuming that damage to 5% of the species of a given ecosystem does not prejudice ecosystem structure and function, HC_5 can be used as criteria for evaluating levels of exposure found in the environment.

The value of HC_5 (3.4 ng/L) indicates that present concentrations of PCBs in open Mediterranean waters, being one order of magnitude or more below the HC_5, should not produce direct adverse effects on the aquatic biological community.

3.4.6.1.2. HCB. Levels of HCB in Mediterranean waters probably do not differ from those found in the Atlantic, due to the very high soil to air and water to air mobility of this chemical. HCB levels in Atlantic waters are of the order of a few picograms per liter. Unlike 2,3,7,8-TCDD or the tributyltin ion (TBT^+), hexachlorobenzene does not have a very high biological activity; thus, direct adverse effects on aquatic organisms are unlikely. Problems can arise from bioconcentration and bioaccumulation phenomena, but these will be discussed in the following chapters.

3.4.6.1.3. HCHs. In the HCH group, α-HCH is characterized by biological activity lower than that of the γ-isomer: at the maximum solubility of 1.4 mg/L, the former had no effect over 2-day experiments on two species of marine phytoplankton (Canton et al., 1978). Very little information is available on the toxicity of ß-HCH. Selected data on NOECs for γ-HCH are shown in Table 3.6.

The remarkable differences in sensitivity to this toxic substance among different species and *phyla* led to a low value of HC_5: 0.094 ng/L (the HC_p approach tends to underestimate the hazardous concentrations when data are in small

Table 3.6. No-Observed-Effect Concentrations (NOECs) of γ-HCH for Representative Marine Organisms and Hazardous Concentration 5% (HC_5)

Organism	NOEC[a] (ng/L)
Microalgae	500,000
Crustacean	100
Bivalve mollusc	100,000
Fish	1000

HC_5 = 0.094 ng/L

[a] Data estimated from Portmann, 1979.

number and quite scattered, as in this case). The obtained HC_5 value is lower than 0.7 ng/L, the concentration measured in Mediterranean waters. This corresponds to protection of about 92% of the aquatic species. Considering all the approximations made to obtain these estimates, this does not necessarily mean that the Mediterranean system is suffering damage from HCH contamination. However, more attention should be paid to this chemical in the future, owing to its variable toxic potential to different organisms.

3.4.6.1.4. DDTs. Most of the available information on the toxicity of these compounds concerns *p,p'*-DDT, although it is well known that *p,p'*-DDE and DDD are the main derivatives of this insecticide. Since DDT seems to be the most biologically active of the three, it will be used for the assessment. Selected NOEC values are shown in Table 3.7. The HC_5 value (13 ng/L) indicates that adverse effects on the aquatic community are unlikely at a level of 0.05 ng/L, corresponding to a protection of more than 99.7% of the species.

3.4.6.2. Marine Mammals and Birds

From the above, it is evident that the only chemical requiring further study of its possible effects on certain species of the Mediterranean aquatic community is γ-HCH. However, the Mediterranean system also includes marine mammals and birds, for which the main mode of exposure is the intake of contaminated food. It is well known that compounds like PCBs, DDTs, and HCB are stable and nonpolar, giving them a high potential for biomagnification, particularly in terrestrial systems and in fish-bird and fish-mammal food chains.

3.4.6.2.1. Marine Mammals. The higher food concentrations of PCBs and DDTs, with respect to HCHs and HCB, are reflected in the tissues of marine mammals. The long life span and low metabolic potency of these animals gives these substances a high accumulation potential. Besides, due to the high fat content of milk, large quantities of SOCs are passed to the newborn from a very early age (Tanabe et al., 1984). As shown by Aguilar (1985), levels of organochlorines in cetacean tissues are proportional to the fat content (particularly triglycer-

Table 3.7 No-Observed-Effect Concentrations (NOECs) of *p,p′*-DDT for Representative Marine Organisms and Hazardous Concentration 5% (HC_5)

Organism	NOEC (ng/L)	Ref.
Microalgae	25,000	Mosser et al., 1972
Crustacean	250	Poole and Willis, 1970
Bivalve mollusc	5000	Butler, 1963
Fish	10,000	Dill and Saunders, 1974
HC_5 = 13 ng/L		

Table 3.8. Chlorinated Hydrocarbons in Subcutaneous Tissues of 9 Finback Whales (*Balaenoptera physalus*) from the Ligurian Sea (Summer 1990)

Chemical	Mean (mg/kg fat)	S.D.	Range
o,p′-DDE	0.07	0.06	0.02–0.20
p,p′–DDE	6.52	4.32	3.19–14.93
o,p′–DDD	0.11	0.10	0.03–0.31
p,p′–DDD	0.90	0.78	0.39–2.57
o,p′–DDT	1.11	1.02	0.39–3.44
p,p′–DDT	1.41	0.98	0.55–3.21
DDTs	10.12	7.23	4.72–24.64
PCBs	6.14	4.18	2.25–14.39

Modified from Focardi et al., 1991.

ides and nonesterified fatty acids): calculating the levels of DDTs and PCBs on a lipid basis, the content in blubber with respect to other tissues comes close to 1:1, with the exception of the brain tissues where the ratio is closer to 10:1 due to the high concentration of the more polar phospholipids. Organs with important metabolic functions (like the liver) may show a higher proportion of degraded forms. Levels in subcutaneous tissues of finback whales *Balaenoptera physalus* encountered in the Ligurian Sea in the summer of 1990 are reported in Table 3.8 in milligrams per kilogram fat. Concentrations are of the order of 10 mg/kg fat for both DDTs and PCBs. Higher concentrations were found in specimens of two dolphin species stranded along the Tyrrhenian coast (1987 to 1989) by Focardi et al. (1990); the data is shown in Table 3.9. Concentrations of PCBs and DDTs found in the blubber of dolphins are about 200 and 100 mg/kg fat. HCB levels are lower. Among DDTs, *p,p′*-DDE accounts for more than 50% of the total residues. About 50% of the PCBs found in both species consisted of the two isomers 2,2′,4,4′,5,5′ and 2,2′,3,4,4′,5′, IUPAC nos. 153 and 138, respectively (Ballschmiter and Zell, 1980).

If the average levels of SOCs in fish are expressed on a fat basis, for a fish containing 5% fat on a wet weight basis, the following levels are obtained for a typical Mediterranean fish:

Table 3.9. Average Concentrations (mg kg^{-1} fat) of Some Chlorinated Hydrocarbons in the Blubber of Dolphins Stranded Along the Tyrrhenian Coast (1987–1989). A: *Stenella coeruleoalba*; B: *Tursiops truncatus*.

	n	HCB	*p,p'*-DDE	*p,p'*-DDD	*p,p'*-DDT	*o,p'*-DDT	PCBs[a]
A	5	0.19	66	5.8	33	18	178
B	2	0.64	52	5.1	11	11	203

[a] As sum of the following (in decreasing order) major components, indicated by IUPAC number (according to Ballschmiter and Zell, 1980): 153, 180, 138 and 170, 187, 196.

From Focardi et al., 1990.

PCBs: 68 ng/g f.w. 5% fat → 1.36 mg/kg fat

HCB: 0.3 ng/g f.w. 5% fat → 0.006 mg/kg fat

DDT: 10 ng/g f.w. 5% fat → 0.2 mg/kg fat.

This means that the enrichment factor dolphin/fish is about 100 (see Table 3.9).

The lower enrichment factor found in finback whales is probably due to a different diet which is less fatty than the diet of fish-eating mammals.

As far as HCHs are concerned, they seem to be degraded rather than accumulated in marine mammals.

3.4.6.2.2. Marine Birds. Among marine birds, there are species like Cory's shearwater *Calonectris diomedea* which feed essentially on fish. Renzoni et al. (1986) measured HCB, *p,p'*-DDE and PCB concentrations, in 15 specimens of this species collected in Majorca, Linosa, and Crete from 1982 to 1984. Average levels of PCBs in fat ranged from 46 to 430 mg/kg fat; *p,p'*-DDE from 28 to 89 mg/kg fat; and HCB from 0.1 to 0.5 mg/kg fat. Audouin's gull *Larus audouinii* is a rare species of marine bird endemic to the Mediterranean and feeding essentially on fish. Eggs of this bird, collected on Capraia Island in the period 1981 to 1986, were used to measure levels of HCB, *p,p'*-DDE, and PCBs (Leonzio et al., 1989). Since the fat content of these eggs is about 1/3 of the dry weight, the measured concentrations are presented in Table 3.10 after conversion to milligrams per kilogram fat.

No significant trends were observed, and the enrichment factor bird/fish calculated from eggs is practically the same as that found in Cory's shearwater fat: about 100, similar to that of dolphins for PCBs, HCB, and DDTs. Comparable results were found by Focardi et al. (1988) in eggs of the yellow-legged herring gull *Larus cachinnans* from Capraia Island (1981 to 1986) where the average concentrations of fat ranged from 0.3 to 0.7 mg/kg for HCB, 18 to 29 mg/kg for *p,p'*-DDE, and 91 to 168 mg/kg for PCBs. In these samples, γ-HCH was measured and the average levels found were 0.15 to 0.27 mg/kg fat, indicating that HCHs

Table 3.10. Levels (mg/kg fat) of Some Organochlorine Compounds in Eggs of Audouin's Gull Collected on Capraia Island (Northern Tyrrhenian) in 1981–1986. Mean Values; n = Number of Samples Analyzed.

Chemical	1981	1982	1983	1984	1985	1986
HCB	0.18	0.12	0.18	0.24	0.09	0.06
p,p'-DDE	20	17	21	19	29	27
PCBs	100	103	124	146	98	86
n	6	23	13	11	12	13

Modified from Leonzio et al., 1989. With permission.

Table 3.11. ADI and Daily Intake in Mediterranean Man with High Fish Consumption (150 g/d)

Chemical	Conc. in fish (ng/g f.w.)	ADI[a] (ng)	Daily intake as % of ADI
PCBs	68	200,000	5.1
HCB	0.3	42,000	0.11
HCHs	0.3	700,000	0.006
DDTs	10	350,000	0.43

[a] ADIs are normalized for a body weight of 70 kg. From UNEP/FAO, 1990.

are metabolized more than other organochlorine compounds (bird/fish enrichment factor about 5). These levels in marine birds are similar to those found by other authors in similar studies in the Mediterranean (Bourne and Bogan, 1980).

3.4.7. Risk Assessment to Man

The data in Table 3.11 is obtained from mean levels in Mediterranean fish on the basis of acceptable daily intake (ADI; FAO/WHO, 1974). The contribution to the ADI of fish consumption is negligible for HCHs and small for HCB; for the two other groups of contaminants it ranges from 0.4 to 5% of the ADI. However, the ADI approach alone may only be appropriate for chemicals having a toxicity threshold: in fact, ADI is defined as "the amount of a substance that can be taken daily in the diet, even over a lifetime, without risk" (FAO/WHO, 1974) and is obtained by applying an arbitrary safety factor to a given NOEL (no-observed-effect level) of exposure. Safety factors cannot be applied to nonthreshold hazardous chemicals (Ramamoorthy and Baddaloo, 1991), as for carcinogens.

In the case of the SOCs discussed here, there is insufficient evidence of human carcinogenicity, but all have shown evidence of carcinogenicity in laboratory animals. It is therefore prudent to regard the substances as presenting carcinogenic risk to humans (UNEP/FAO, 1990; IARC, 1987). Consequently, a zero-threshold is assumed, with a response/dose relationship leading to increased cancer incidence with increasing exposure.

Animal experiments are currently being carried out at high doses, so that risk evaluation based on these data requires both high-dose to low-dose extrapolation

Table 3.12. Carcinogenic Potency of Some SOCs in Humans, as "upper-bound slope", B_H, and daily intake, I, Associated with 10^{-5} Lifetime Risk

chemical	Human upper bound slope (B_H)[a] (kg·d/mg)	I (mg/d)
PCBs	4.43	1.58×10^{-4}
HCB	1.68	4.17×10^{-4}
γ-HCH	1.33	5.26×10^{-4}
DDT	8.42	8.31×10^{-5}

[a] From Anderson, 1985; 1988. With permission.

and species-to-species extrapolation. The linearized multistage model (WHO, 1987) provides an estimate of the carcinogenic potency of certain chemicals in humans, expressed as "upper-bound slope" (Table 3.12; Anderson, 1985; 1988).

The intake rate I (in mg/d) associated with a one in 100,000 (i.e., 10^{-5}) lifetime risk in humans is calculated as follows.

$$I = 70 \times 10^{-5} / B_H$$

where 70 is the body weight (kg), 10^{-5} the lifetime risk, and B_H the human upper bound slope (kg d/mg) (Stara et al.,1987).

Given SOC levels in fish, the daily intake associated with 1, 3, and 7 fish meals (150 g each) per week can be calculated (Table 3.13). Cancer risk is assumed to be directly proportional to the intake of a carcinogenic substance. If so, for a given dose, the proportion of people who would develop cancer after life-long exposure can be calculated. Table 3.14 shows the proportion of expected increase in cancer risk due to life-long exposure to present levels of organochlorine compounds from Mediterranean fish consumption.

PCBs and DDT show an expected increase in cancer incidence higher than 10^{-5}, even in the one meal per week consumers. This exceeds the "acceptable" risk indicated by WHO (1987), which is 10^{-5}, or one cancer per 100,000 people after life-long exposure.

3.4.8. Criteria for Environmental Recovery

Ecotoxicology, the science for the study of the environmental fate and effects of contaminants, is still young (Truhaut, 1975; Butler, 1978; Moriarty, 1983; Calamari and Vighi, 1990). As a consequence, uncertainties in final evaluations may be relatively high, as in the case of the use of mathematical models for extrapolating cancer incidence in laboratory animals from high to low doses, and then from animals to humans (Kimbrough, 1990). Besides, available tools to evaluate the effects in ecosystems are still rough and incomplete: the main limitation appears to be an excess of "anthropocentrism" in environmental quality criteria where the need to protect humans is too many times disconnected from the need to protect the natural system. However, from the available information discussed here, the following conclusions can be drawn:

Table 3.13. Daily Intakes Associated with Fish Consumption: 1, 3 and 7 Meals per Week

		Weekly		
Chemical	Conc. in fish (ng/g) (wet weight)	1 Meal (mg/d)	3 Meals (mg/d)	7 Meals (mg/d)
PCBs	68	1.46×10^{-3}	4.37×10^{-3}	1.02×10^{-2}
HCB	0.3	6.43×10^{-6}	1.93×10^{-5}	4.50×10^{-5}
γ-HCH	0.3	6.43×10^{-6}	1.93×10^{-5}	4.50×10^{-5}
DDT	10	2.14×10^{-4}	4.29×10^{-4}	1.50×10^{-3}

Table 3.14. Lifetime Cancer Risk to Humans from Exposure to Selected SOCs from Mediterranean Fish Consumption

		Cancer risk		
Chemical	Conc. in fish (ng/g, f.w.)	1 Meal weekly	3 Meals weekly	7 Meals weekly
PCBs	68	9.2×10^{-5}	2.8×10^{-4}	6.5×10^{-4}
HCB	0.3	1.5×10^{-7}	4.6×10^{-7}	1.1×10^{-6}
γ-HCH	0.3	1.2×10^{-7}	3.7×10^{-7}	8.6×10^{-7}
DDT	10	2.6×10^{-5}	7.7×10^{-5}	1.8×10^{-4}

- High quantities, in decreasing order, of PCBs, HCHs, and DDTs are present in air, water, sediment, and organism compartments of the Mediterranean Sea system: 5000, 3000, and 1000 metric tons respectively.
- The Mediterranean quantity of HCB is still limited (50 t), probably due to its lower production rate; in addition, the high partition in air (two orders of magnitude higher than other SOCs) may lead to faster distribution (and dilution) of the chemical in the biosphere; due to its high bioconcentration and biomagnification potential, levels of HCB in marine organisms are relatively high, particularly in fish-eating mammals and birds.
- Water is the main reservoir of HCHs (>99% of the total), while HCB, DDTs, and PCBs are mainly stored in sediments (54–90%).
- Marine mammals and birds have a high exposure to PCBs, HCB, and DDTs from fish eating. A magnification factor of 100 exists between the levels in these animals and fish after normalization of concentration data on a fat basis. Due to metabolic processes, HCHs are not accumulated to the same extent in marine mammals and birds.
- HCHs exceed the hazardous concentration 5% (HC_5) level for the biological community of Mediterranean marine organisms by a factor of 10, indicating the possibility of a change in marine ecosystem structure and functions.
- PCBs and DDTs may cause a cancer incidence exceeding 1/100,000 (risk > 10^{-5}) in human consumers of fish after life-long exposure.

As previously discussed, these contaminants travel as vapors from land-based sources, mostly by air transport. Sources include all, even "clean", soils receiving from and releasing to the air SOC vapors. These aspects are now well known, as illustrated in the recent book edited by Kurtz (1990). Air/water partition is such that the Mediterranean receives contamination from its basin, as well as from other parts of the world. At the same time, the water-to-air flows of SOCs mean that the Mediterranean releases eventual excesses of contaminants, displacing the problem elsewhere. This process is less effective for deep-sea environments (below 200 m), particularly for contaminants such as PCBs, HCB, and DDTs which are accumulated in sediments. If contaminant input to the Mediterranean were to stop, sediments would behave as a secondary source, retarding recovery for a long time.

In this analysis of the Mediterranean system, the terrestrial environment around it was not included. However, the differences in environmental loads, essentially in the terrestrial environment, and quantities of SOCs found in the Mediterranean (excluding land) indicate that terrestrial soils constitute a large SOC reservoir (Gaggi et al., 1985). These are not necessarily soils of Mediterranean countries. Thus, the theoretical solution to the problem is to completely phase out production of these substances on a worldwide basis.

Control measures to mitigate the environmental effects of contaminants can be taken at three main levels. The first is regulation of production; the second is regulation of utilization and disposal; and the third is to establish levels of contamination in environmental matrices (e.g., water) that are legally acceptable and considered as harmless (e.g., *environmental quality objective*). When facing stable and sufficiently mobile contaminants, the environmental quality objective approach may not be appropriate.

Local or single-country actions, even drastically reducing or phasing out the input of these substances into the environment, will not be completely successful unless similar action is taken on a global scale. In the last 20 years, some technically advanced countries have greatly reduced the use of refractory chemicals such as PCBs and DDT, obtaining only partial environmental recovery. This is due to the long time necessary to deplete environmental reservoirs (such as soils and sediments) and to transfrontier chemical fallout from countries where these substances are still in use. Control of the use is therefore an intermediate objective, the final one being the control of production.

3.5. UNCERTAINTY OF PRESENT APPROACHES: FUTURE RESEARCH NEEDS

The approaches in evaluating hazard and risk associated with the environmental contamination by organic chemicals are based on the knowledge of environmental fate and on the measurements of effects due to known exposure.

The investigation of the environmental fate of contaminants may receive great impetus from the approach of evaluative models. These are based on the interconnections existing between the molecular and molar properties of the chemicals (instrinsic properties) and their behavior in the environment. In this way, *after the DDT is known, all other chemicals with similar intrinsic properties are expected to behave similarly (the DDT-like compounds, such as some of the polychlorinated dioxins).* However, environmental variables (properties of the system) may lead to different results under different conditions: the high temperature of soil in equatorial countries enhances the dissipation by volatilization while in cold areas or on the top of high mountains, deposition of air-borne contaminants is favored. Besides, a good knowledge of possible products and metabolites originating from a mother compound is necessary. In past years, the availability of data on physicochemical properties of contaminants has increased, as has the number and accuracy of property estimation methods (Lyman et al., 1990). Site-specific situations, like the high vulnerability of a water table due to a high soil permeability, may require special attention. In general, considering the increasing standardization of human lifestyles, experiences with different chemicals in different environments may be applied to predict the impact in new sites. Increasing the number of known chemicals means improving present predictive capability.

In the case of measuring toxicity, information is generally available for acute toxicity, somewhat less for chronic toxicity, and still quite limited for carcinogens. Recent approaches are trying to correlate short-term mutagenicity test results with long-term carcinogenicity rodent studies. For noncarcinogenic chemicals, the problem of statistical inference and species-to-species extrapolation have traditionally been solved by arbitrary or calculated safety factors (1/10 to 1/1000). Some esperimental approaches need to be reconsidered in order to reduce the probability of false negative findings.

The main problem in evaluating carcinogenic potency is the uncertainty due to:

- The assumption that the little group of observed laboratory organisms represent a larger population, if not *the species* (statistical inference)
- High-dose to low-dose extrapolations
- Interspecies extrapolations
- Intraspecies variability

Notwithstanding these problems, the risks calculated by current methods, such as the upper-bound slope for humans, B_H, are largely conservative, in the sense that they overestimate the actual risk and can be considered adequately precautionary.

The major limitation of present ecotoxicological approaches consists in the small number of instruments available for evaluating effects at the ecosystem level. Recently, advances have been made in this field, pointing out some basic concepts such as those suggested by Petersen and Petersen (1989):

- Natural population mortality should be considered: possible compensator mechanisms may mask the effect of toxicants.
- Species with broad ecological niches (generalists) will be more resistant than species with narrow niches.
- *Keystone species* need to be identified; an adverse effect on these may produce dramatic effects on the diversity and structure of the biological community (keystone species have a particular role in controlling the structure of a community, like the keystone in the Roman arch).

Another crucial aspect pointed out by the same authors (Petersen and Petersen, 1989) is that *structural properties of communities are less conservative than functional properties*. In other words, a change in structure (i.e., a modification in species composition) tends to occur *before* changes in function.

The HC_p approach is fascinating and, with some adaptation to different systems by varying the p value adequately, may help to evaluate effects at the ecosystem level, particularly if some intrinsic limitations of NOEL are overcome.

The possibility of investigating long-term effects on keystone species, directly in the field and with an appropriate time scale, should be considered.

A final consideration: the main problem in protecting environmental quality is that, still now, it depends too much on anthropocentric criteria. The case of marine coastal waters is typical: every year, during the period of recreational use of the sea (summer time in the Mediterranean countries) there is an enormous increase in public concern on their quality. Almost every day mass media provide information on the "health" of coastal waters. Or, better, on the potential effects to humans exposed to these waters by swimming (and then drinking small quantities of them). One of the most widely applied criteria is based on the detection of fecal coliform bacteria, as tracers of untreated sewage. The approach is quite correct to provide an answer to the question "What will happen to humans after swimming in those waters?". The problem arises from the incorrect use of results when nonexpert people are left to believe that, by the coliform bacteria test, we are able to answer the question "Are these waters clean?" or, even worse, "Is the marine ecosystem damaged?". The absence or presence of fecal bacteria is not related to the conditions of marine environments. From this, the need for simple, feasible, and reliable tests to measure the quality of waters not only for swimming humans, but also for swimming and non swimming marine organisms. For people working in applied ecology and ecotoxicology, it is very hard to accept that the only problems in marine environments are those affecting human health directly, as in the case of bathing waters, or indirectly, as in the case of contamination of seafood and fishery products. All these aspects are important, but a little bit too partial: for man's survival, a navigable ocean is not enough!

3.6. CONCLUSIONS

New tools are continuously produced to improve the present capability of understanding the environmental behavior of chemicals: physicochemical data of pure substances are progressively refined and site-specific studies on contamination and pollution cases improve our predictive ability. There is the possibility that relative results obtained by evaluative models (*rankings*) could be calibrated under specific field conditions, leading to predictions of absolute data with acceptable approximation. The quality of toxicological data is improving and mechanisms of action have become clearer. Present limitations due to an excess of anthropocentrism in toxicological criteria indicate the need for an effort in the direction of the environment. What still seems to be weak is the voice of new ecotoxicologists, able to suggest new approaches and take advantage of all the instruments, old and new, produced by basic research (*the stepping-stones*). In the near future, there will be an increasing need for people able to cross the stream of hazard assessment and risk analysis by walking on the stepping-stones, with an appropriate equilibrium and a little bit of art.

REFERENCES

Abbott D.C., Harrison, R.B., Tatton, J. O'G., and Thomson, J., Organochlorine pesticides in the atmospheric environment, *Nature*, 208: 1317-1318 (1965).

Aguilar, A., Compartmentation and reliability of sampling procedures in organochlorine pollution surveys of Cetaceans, *Residue Rev.*, 95: 91-114 (1985).

Anderson, E.L., Quantitative approaches in use in the United States to assess cancer risk, in *Methods of Estimating Risk of Chemical Injury: Human and Non-human Biota and Ecosystem.* SCOPE 26, SGOMSEC 2, Vouk, V.B., Butler, G.C., Hoel, D.G., and Peakall, D.B., Eds. John Wiley & Sons, Chichester, U.K., pp. 405-436 (1985).

Anderson, E.L., The risk analysis process, in *Carcinogen Risk Assessment*, Travis, C.C., Ed. Plenum Press, New York, pp. 3-17 (1988).

Anonymous, Report of a new chemical hazard, *New Scientist*, Dec. 15, 1966, p. 612 (1966).

Atlas, E.L., and Schauffler, S., Concentration and variation of trace organic compounds in the North Pacific atmosphere, in *Long-Range Transport of Pesticides*, Kurtz, D.A., Ed. Lewis Publishers, Chelsea, MI, pp. 161-183 (1990).

Bacci, E., *Semivolatile Organochlorinated Hydrocarbons in the Mediterranean Sea: Sources, Fate, Hazard Assessment and Proposed Control Measures*, UNEP/MAP Agreement 042/90, Athens, Greece (1991).

Bacci, E., and Calamari, D., Learning from field work and modelling, *Proceedings of the Workshop on Chemical Exposure Prediction*, Trois Epis (Colmar, France), June 11-13, 1990. European Science Foundation, Strasbourg, France, pp. 198-208 (1990).

Bacci, E., Calamari, D., Gaggi, C., and Vighi, M., Bioconcentration of organic chemical vapors in plant leaves: experimental measurements and correlation, *Environ. Sci. Technol.*, 24: 885-889 (1990a).

Bacci E., Cerejeira, M.J., Gaggi, C., Chemello, G., Calamari, D., and Vighi, M., Bioconcentration of organic chemical vapours in plant leaves: the azalea model, *Chemosphere*, 21: 525-535 (1990b).

Bacci, E., Cerejeira, M.J., Gaggi, C., Chemello, G., Calamari, D., and Vighi, M., Chlorinated dioxins: volatilization from soils and bioconcentration in plant leaves, *Bull. Environ. Contam. Toxicol.*, 48: 401-408 (1992).

Ballschmiter, K., and Zell, M., Analysis of polychlorinated biphenyls (PCB) by glass capillary gas chromatography, *Fresenius Z. Analyt. Chem.*, 302: 20-31 (1980).

Bidleman, T.F., Billings, W.N., and Foreman, W.T., Vapor-particle partitioning of semivolatile organic compounds: estimates from field collections, *Environ. Sci. Technol.*, 20: 1038-1043 (1986).

Bidleman, T.F., Patton, G.W., Hinckley, D.A., Walla, M.D., Cotham, W.E., and Hargrave, B.T., Chlorinated pesticides and polychlorinated biphenyls in the atmosphere of the Canadian Arctic, in *Long-Range Transport of Pesticides,* Kurtz, D.A., Ed. Lewis Publishers, Inc., Chelsea, MI, pp. 347-372 (1990).

Boon, J.P., van Zantvoort, M.B., Govaert, M.J.M.A., and Duinker, J.C., Organochlorines in benthic polychaetes (*Nephtys* spp.) and sediments from the Southern North Sea. Identification of individual PCB components, *Netherland J. Sea Res.*, 19: 93-109 (1985).

Bourne, W.R.P., and Bogan, J.A., Organochlorines in Mediterranean seabirds, *Environ. Conserv.*, 7: 277-278 (1980).

Burton, M.A.S., and Bennet, B.G., Exposure of man to environmental hexachlorobenzene (HCB) - An exposure committment assessment, *Sci. Tot. Environ.*, 66: 137-146 (1987).

Butler, G.C., Ed., *Principles of Ecotoxicology*, SCOPE 12, John Wiley & Sons, New York, (1978).

Butler, P.A., *Commercial Fishery Investigations*. U.S. Dept. Int. Fish and Wildlife Serv. Circ. 167 (1963).

Calamari, D., Bacci, E., Focardi, S., Gaggi, C., Morosini, M., and Vighi, M., The role of plant biomass in the global partition of chlorinated hydrocarbons, *Environ. Sci. Technol.*, 25: 1489-1495 (1991).

Calamari, D., and Vighi, M., The role of Ecotoxicology in environmental protection, in *Basic Science in Toxicology*, Volans, G.N., Sims, J., Sullivan, F.M., and Turner, P., Eds. Taylor & Francis, London, UK, pp. 193-206 (1990).

Canton, J.H., Wegman, R.C.C., Vulto, T.J.A., Verhoff, C.H., and van Esch, G.J., Toxicity, accumulation and elimination studies of γ-hexachlorocyclohexane (γ-HCH) with saltwater organisms of different trophic levels, *Water Res.*, 12: 687-690 (1978).

Carson, R.L., *The Sea Around Us*, Oxford University Press, New York, (1950).

Cole, H., Barry, D., Frear, D.E.H., and Bradford, A., DDT levels in fish, streams, stream sediments, and soil before and after aerial spray application for fall cankerworm in Northern Pennsylvania, *Bull. Environ. Contam. Toxicol.*, 2: 127-146 (1967).

Dalla Venezia, L., and Fossato, V.U., Effects of PCBs on *Leander adspersus*: toxicity, bioaccumulation, oxygen consumption, osmoregulation, *FAO Fish. Rep. 334 Suppl.*: 39-49 (1986).

DeFoe, D.L., Veith, G.D., and Carlson, R.W., Effects of Aroclor 1248 and 1260 on the fathead minnow (*Pimephales promelas*), *J. Fish. Res. Board Can.*, 35: 997-1002 (1978).

Dill, P.A., and Saunders, R.C., Retarded behavioral development and impaired balance in atlantic salmon (*Salmo salar*) alevins hatched from gastrulae exposed to DDT, *J. Fish. Res. Board Can.*, 31: 1936-1938 (1974).

FAO/WHO, Toxicological evaluation of certain food additives in a review of general principles and specifications, *Tech. Rep. Ser. No. 539*, WHO, Geneva (1974).

Fisher, N.S., and Würster, C.F., Individual and combined effects of temperature and polychlorinated biphenyls on the growth of three species of phytoplankton, *Environ. Pollut.*, 5: 205-212 (1973).

Focardi, S., Fossi, C., Lambertini, M., Leonzio, C., and Massi, A.. Long term monitoring of pollutants in eggs of yellow-legged herring gull from Capraia Island (Tuscan Archipelago), *Environ. Monitoring Assessment*, 10: 43-50 (1988).

Focardi, S., Marsili, L., Fabbri, F., and Carlini, R., Preliminary study of chlorinated hydrocarbon level in *Cetacea* stranded along the Tyrrhenian coast of Latium (Central Italy), in *European Research on Cetaceans*. 4. Proceedings of the 4th Annual Conference of the European Cetacean Society, Evans, P.G.H., Aguilar, A., and Smeenk, C., Eds. Palma de Mallorca, Spain, 2-4 March 1990, pp. 108-110 (1990).

Focardi, S., Notarbartolo di Sciara, G., Venturino, C., Zanardelli, M., and Marsili, L., Subcutaneous organochlorine levels in finback whales (*Balaenoptera physalus*) from the Ligurian Sea, in *European Research on Cetaceans*. 5. Proceedings of the European Cetacean Society, Evans, P.G.H., Ed. Sandefjord, Norway, 21-23 February 1991, pp. 93-96 (1991).

Foreman, W.T., and Bidleman, T.F., An experimental system for investigating vapor-particle partitioning of trace organic pollutants, *Environ. Sci. Technol.*, 21: 869-875 (1987).

Gaggi, C., Bacci, E., Calamari, D., and Fanelli, R., Chlorinated hydrocarbons in plant foliage: an indication of the tropospheric contamination level, *Chemosphere*, 14: 1673-1686 (1985).

GESAMP, Atmospheric input of trace species to the world ocean. GESAMP Working Group No. 14, XIX/4, March 1989, WMO, Athens (1989).

Geyer, H., Freitag, D., and Korte, F., Polychlorinated biphenyls (PCBs) in the marine environment, particularly in the Mediterranean, *Ecotoxicol. Environ. Saf.*, 8: 129-151 (1984).

Goldstein, J.A., Structure-activity relationships for the biochemical effects and the relationship to toxicity, in *Halogenated Biphenyls, Terphenyls, Naphthalenes, Dibenzodioxins and Related Products*, Kimbrough, R.D., Ed. Elsevier/North-Holland, Biom. Press, Amsterdam, The Netherlands, pp. 151-190 (1980).

Guicherit, R., and Schulting, F.L., The occurrence of organic chemicals in the atmosphere of the Netherlands, *Sci. Tot. Environ.*, 43: 193-219 (1985).

Hammond, E.C., Selikoff, I.V., and Seidman, H., Asbestos exposure, cigarette smoking and death rates, *Ann. N.Y. Acad. Sci.*, 330: 473-490 (1979).

Hansen, L.G., Environmental toxicology of polychlorinated biphenyls, in *Polychlorinated Biphenyls (PCBs): Mammalian and Environmental Toxicology*, Safe, S., Ed. Springer-Verlag, Berlin, pp. 15-48 (1987).

Heinisch, E., Input of pesticide agents from non-agricultural sources into environment, *Ekològia (CSSR)*, 4: 97-109 (1985).

Hidaka, H., and Tatsukawa, R., Environmental pollution of chlorinated hydrocarbons around Syowa station, *Antarctic Rec.*, 80: 14-29 (1983).

Hinckley, D.A., Bidleman, T.F., and Rice, C.P., Atmospheric organochlorine pollutants and air-sea exchange of hexachlorocyclohexane in the Bering and Chukchi Seas, *J. Geophys. Res.*, 96: 7201-7213 (1991).

Hutzinger, O., Safe, S., and Zitko, V., *The Chemistry of PCB's*. CRC Press, Boca Raton, FL, (1974).

Karickhoff, S.W., Brown, D.S., and Scott, T.A., Sorption of hydrophobic pollutants on natural sediments, *Water Res.,* 13: 241-248 (1979).

Kimbrough, R.D. Environmental protection: theory and practice, *Environ. Sci. Technol.*, 24: 1442-1445 (1990).

Krämer, W., and Ballschmiter, K., Global baseline studies XII. Content and pattern of polychloro-cyclohexanes (HCH) and -biphenyls (PCB), and content of hexachlorobenzene in the water column of the Atlantic Ocean, *Fresenius Z. Anal. Chem.*, 330: 524-526 (1988).

Kurtz, D.A., Ed. *Long-Range Transport of Pesticides* Lewis Publishers, Chelsea, MI (1990).

Kurtz, D.A., and Atlas, E.L., Distribution of hexachlorocyclohexanes in the Pacific Ocean basin, air and water, 1987, in *Long range transport of pesticides*, Kurtz, D.A., Ed. Lewis Publishers, Inc., Chelsea, MI, pp. 143-160 (1990).

IARC, *Overall Evaluations of Carcinogenicity; An Updating of IARC Monographs Volumes 1-42. Supplement 7.* WHO/IARC (1987).

Leonzio, C., Lambertini, M., Massi, A., Focardi, S., Fossi, C., An assessment of pollutants in eggs of Audouin's gull (*Larus audouinii*), a rare species of the Mediterranean Sea, *Sci. Tot. Environ.*, 78: 13-22 (1989).

Mackay, D., Finding fugacity feasible, *Environ. Sci. Technol.*, 13: 1218-1223 (1979).

Mackay, D., Correlation of bioconcentration factors, *Environ. Sci. Technol.*, 16: 274-278 (1982).

Mackay, D., and Paterson, S., Fugacity revisited, *Environ. Sci. Technol.*, 16: 654-660 (1982).

Mackay D., Paterson, S., and Schroeder, W.H., Model describing the rates of transfer processes of organic chemicals between the atmosphere and water, *Environ. Sci. Technol.*, 20: 810-816 (1986).

McConnell, E.E., and McKinney, J.D., Exquisite toxicity in the Guinea pig to structurally similar halogenated dioxins, furans, biphenyls and naphthalenes, *Toxicol. Appl. Pharmacol.*, 45: 298 (Abstr., 1978).

Metcalf, R.L., A century of DDT, *J. Agric. Food Chem.*, 21: 511-519 (1973).

Miller, S. The persistent PCB problem, *Environ. Sci. Technol.*, 16: 98A-99A (1982).

Moriarty, F., *Ecotoxicology. The Study of Pollutants in Ecosystems,* Academic Press, London, (1983).

Morrison, F., A review of the use and place of lindane in the protection of stored products from the ravages of insect pests, *Residue Rev.,* 41: 113-180 (1972).

Mosser, J.L., Fisher, N.S., and Wurster, C.F., Polychlorinated biphenyls and DDT alter species composition in mixed cultures of algae, *Science*, 176: 533-535 (1972).

Nimmo D.R., Hanson, R.J., Couch, J.A., Cooley, N.R., Parrish, P.R., and Lowe, J.I., Toxicity of aroclor 1254 and its physiological activity in several estuarine organisms, *Arch. Environ. Cont. Toxicol.*, 3: 22-29 (1975).

OECD, *Report of the OECD Workshop on Ecological Effects Assessment*. Organization for Economic Co-operation and Development, Paris, France (1989).

Peattie, M.E., Lindsay, D.G., and Hoodless, R.A., Dietary exposure of man to chlorinated benzenes in the United Kingdom, *Sci. Tot. Environ.*, 34: 73-86 (1984).

Perera, F., Biological markers in risk assessment, in *Carcinogen Risk Assessment*, Travis, C.C., Ed. Plenum Press, New York, pp. 123-138 (1988).

Petersen, R.C., Jr., and Petersen, L.B.M., Ecological concepts important for the interpretation of effects of chemicals on aquatic systems, in *Chemicals in the Aquatic Environment. Advanced Hazard Assessment*, Landner, L., Ed. Springer-Verlag, Berlin, pp. 165-196 (1989).

Picer, N., and Picer, M., Monitoring of Chlorinated Hydrocarbons in Water and Sediments of the North Adriatic Coastal Waters, *J. Etud. Pollut. CIESM*, 4: 133-136 (1979).

Poole, R.L., and Willis, M., Effects of some pesticides on larvae of the market crab *Cancer magister* and the red crab *Cancer productus* and the bioassay of industrial wastes with crab larvae, *Calif. Dept. Fish. Games Marine Res. Lab.*, Report No 70/15, 19 pp (1970).

Portmann, J.E., Evaluation of the impact on the aquatic environment of hexachlorocyclohexane (HCH isomers), hexachlorobenzene (HCB), DDT (+DDE and DDD), heptachlor (+heptachlor epoxide) and chlordane, prepared for the Commission of the European Communities—Environment and Consumer Protection Service; pp. 337 (1979).

Quinby, G.E., Hayes, W.J., Armstrong, J.F., and Durham, W.F., DDT storage in U.S. population, *J. Am. Med. Assoc.*, 191: 175-179 (1965).

Ramamoorthy, S., and Baddaloo, E., *Evaluation of Environmental Data for Regulatory and Impact Assessment. Studies in Environmental Science 41*, Elsevier, Amsterdam, (1991).

Renzoni, A., Focardi, S., Fossi, C., Leonzio, C., and Mayol, J., Comparison between concentrations of mercury and other contaminants in eggs and tissues of Cory's shearwater *Calonectris diomedea* collected on Atlantic and Mediterranean Islands, *Environ. Pollut., Ser. A*, 40: 17-35 (1986).

Rippen, G., and Frank, R., Estimation of hexachlorobenzene pathways from the technosphere into the environment, in *Hexachlorobenzene: Proceedings of an International Symposium*, Morris, C.R., and Cabral, J.R.P., Eds. IARC Scientific Publications No. 77, International Agency for Research on Cancer, Lyon, France, pp. 45-52 (1986).

Risebrough, R.W., Beyond long-range transport: a model of a global gas chromatographic system, in *Long-Range Transport of Pesticides*, Kurtz, D.A., Ed. Lewis Publishers, Chelsea, MI, pp. 417-426 (1990).

Risebrough, R.W., Huggett, R.J., Griffin, J.J., and Goldberg, E.D., Pesticides: transAtlantic movements in the northeast trades, *Science*, 159: 1233-1236 (1968).

Sladen, W.J.L., Menzie, C.M., and Reichel, W.L., DDT residues in Adelie penguins and crabeater seal from Antarctica: ecological implications, *Nature*, 210: 670-673 (1966).

Stara, J.F., Bruins, R.J.F., Dourson, M.L., Erdreich, L.S., Hertzberg, R.C., Durkin, P.R., and Pepelko, W.E., Risk assessment is a developing science: approaches to improve evaluation of single chemicals and chemical mixtures, in *Methods for Assessing the Effects of Mixtures of Chemicals*, SCOPE 30, SGOMSEC 3. Vouk, V.B., Butler, G.C., Upton, A.C., Parke, D.V., and Asher, S.C., Eds. John Wiley & Sons, Chichester, U.K., pp. 719-743 (1987).

Stewart, E.H., and Windsor, J.G., Jr., Exposure assessemt of sewage treatment plant effluent by a selected chemical marker method, *Arch. Environ. Contam. Toxicol.*, 19: 674-679 (1990).

Suntio, L.R., Shiu, W.Y., Mackay, D., Seiber, J.N., and Glotfelty, D., Critical review of Henry's law constants for pesticides, *Rev. Environ. Contam. Toxicol.*, 103: 1-59 (1988).

Tanabe, S., Tanaka, H., and Tatsukawa, R., Polychlorinated biphenyls, DDT and hexachlorocyclohexane isomers in the western North Pacific ecosystems, *Arch. Environ. Contam. Toxicol.*, 13: 731-738 (1984).

Tardiff, R.G., and Rodricks, J.V., Comprehensive risk assessment, in *Toxic Substances and Human Risk. Principles of Data Interpretation*, Tardiff, R.G., and Rodricks, J.V., Eds. Plenum Press, New York, pp. 391-434 (1987).

Tatsukawa, R., Wakimoto, T., and Ogawa, T., BHC residues in the environment, in *Environmental Toxicology of Pesticides,* Matsumara, F., Boush, G. M., and Misato, T., Eds. Academic Press, New York, pp. 229-238 (1972).

Thomas, R.G., Volatilization from water, in *Handbook of Chemical Property Estimation Methods*, Lyman, W.J., Reehl, W.F., Rosenblatt, D.H., Eds. American Chemical Society, Washington, D.C., chap. 15, pp. 1-34 (1990).

Truhaut, R., Ecotoxicology—a new branch of toxicology: a general survey of its aims, methods and prospects, in *Ecological Toxicology Research*, McIntyre, A.D., and Mills, C.F., Eds. Plenum Press, New York, pp. 3-23 (1975).

UNEP, *Convention for the Protection of the Mediterranean Sea Against Pollution and Its Related Protocols*, United Nations, New York (1982).

UNEP, *Pollutants from Land-Based Sources in the Mediterranean*, UNEP Regional Seas Reports and Studies No. 32, UNEP, Athens (1984).

UNEP, Co-ordinated Mediterranean pollution monitoring and research programme (MED POL-Phase I). Final Report. 1975-1980. *MAP Tech. Rep. Ser. No. 9,* UNEP, Athens; pp. 276 (1986).

UNEP/FAO, Baseline studies and monitoring of DDT, PCBs and other chlorinated hydrocarbons in marine organisms (MED POL III). *MAP Tech. Rep. Ser. No. 3*, UNEP, Athens; pp. 128 (1986).

UNEP/FAO, Assessment of the state of pollution of the Mediterranean Sea by organohalogen compounds. *MAP Tech. Rep. Ser. No. 39,* UNEP, Athens, pp. 224 (1990).

U.S. EPA, *Guidelines for Exposure Assessment.* Federal Register, 51 No. 185: 34042-34054 (1986).

van Leeuwen, K., Ecotoxicological effects assessment in the Netherlands: recent developments, *Environ. Manage.*, 14: 779-792 (1990).

van Straalen, N.M., and Denneman, C.A.J., Ecotoxicological evaluation of soil quality criteria, *Ecotoxicol. Environ. Safety*, 18: 241-251 (1989).

Villeneuve, J.P., Polychlorinated biphenyls in near-sea atmospheric samples from the Mediterranean in 1975 to 1977, *J. Etud. Pollut. CIESM* 7, (1984): 489-493 (1985).

Villeneuve, J.P., Elder, D.L., and Fukai, R., Distribution of polychlorinated biphenyls in seawater and sediments from the open Mediterranean Sea, *J. Etud. Pollut. CIESM* 5, (1980): 251-256 (1981).

White, A.V., and Burton, I., Eds., *Environmental Risk Assessment*, SCOPE 15, John Wiley & Sons, Chichester, U.K. (1980).

WHO. *Potentially Toxic Microorganic Substances in Drinking-Water, EUR/ICP/CWS 013*, World Health Organization, Geneva (1987).

WHO. DDT and its derivatives—environmental aspects, *Environmental Health Criteria 83*, World Health Organization, Geneva (1989).

Wittlinger, R., and Ballschmiter, K., Global baseline pollution studies XI: congener-specific determination of polyclorinated biphenyls (PCB) and occurrence of α- and γ-hexachlorocyclohexane (HCH), *4,4'*-DDE and *4,4'*-DDT in continental air, *Chemosphere*, 16: 2497-2513 (1987).

Woodwell, G.M., Craig, P.P., and Johnson, H.A., DDT in the biosphere: where does it go? *Science*, 174: 1101-1107 (1971).

Woodwell, G.M., Wurster, C.F., and Isaacson, P.A., DDT residues in an East Coast estuary: a case of biological concentration of a persistent insecticide, *Science*, 156: 821-824 (1967).

APPENDIX 1

Accuracy of log-transformation of exponential and power models

Exponential and power models are largely applied in ecotoxicology both with interpretative and predictive purposes. The log- or ln-transformed variables are linearly correlated and allow use of simple fitting methods, such as the least-squares. So, an exponential relationship such as the following:

$$Y_{(t)} = Y_{(o)} \exp(-bX) \qquad (1)$$

becomes, after ln-transformation:

$$\ln(Y_{(t)}) = \ln(Y_{(o)}) - bX \qquad (2)$$

As recently discussed by Newman (1993; reference list section 1), Equation 2 is incomplete: the random error term, E, is missing. The complete form of Equation 2 is as follows:

$$\ln(Y_{(t)}) = \ln(Y_{(o)}) - bX + E \qquad (3)$$

The random error term has a mean of zero in the logarithmic units, but not in the arithmetic units of the original equation. Consequently, the error term can be omitted in regression based on untransformed variables, while in regressions where the variables are transformed the same term must be retained for back-transformation:

$$Y_{(t)} = Y_{(o)} \exp(-bX) \exp(E) \qquad (4)$$

Assuming a normal distribution of the regression residuals, the random error term E can be obtained as follows:

$$E = MSE/2 \qquad (5)$$

where MSE is the mean square of the error from the regression calculated from the sum of the squared residuals for each data pair, divided by the degrees of freedom (n–2, with n = number of pairs).

The same approach should be applied to power relationships, such as:

$$Y = aX^b \qquad (6)$$

When the parameters a and b are calculated on the log-transformed expression, the random error term, E, must be considered:

$$\log (Y) = \log (a) + b \log (X) + E \tag{7}$$

and reintroduced in the back-transformation as follows:

$$Y = aX^b \cdot 10^E \tag{8}$$

Often E is very little, and the correcting factors exp (E) and 10^E in Equations 4 and 8 are tending to one. When so, they may be neglected. In other cases, it may happen that experimental data are quite rough and a correcting factor of, for instance, 1.08, leading to an increase in the calculated value of dependent variable of 8%, is not practically significant. However, even if in the past it was not the case, at least for future investigations it is important to consider the possibility of the bias generated by logarithmic transformations, improving in this way the accuracy of predictive models.

APPENDIX 2

A simple BASIC program to calculate the Hazardous Concentration to the proportion p of the species of an ecosystem

```
10 print "          Calculating HCp and q"
20 print "according to van Straalen and Denneman, 1989"
30 print "please input the kind of data (e.g. NOEC, LC50)"
40 input A$
50 print "units of" A$
60 input B$
70 print "how many data sets have been selected?"
80 input N
90 print "please enter the value of p (e.g. 0.05)"
100 input P
110 dim I(N)
120 dim L1(N)
130 dim L2(N)
140 for I=0 to (N-1)
150 print "please input the value of "A$
160 input I(N)
170 L1(N)=LOG(I(N))
180 L2/N)=(L1(N))^2
190 S1=S1+L1(N)
200 S2=S2+L2(N)
210 NEXT I
220 AV=S1/N
230 rem geometric mean of A$ = exp(AV) = AG
240 AG=EXP(AV)
250 rem SM = standard deviation of ln A$
260 SM=((S2-((S1)^2)/N)/(N-1))^(1/2)
270 print "please input Kooijman's dm value (see Table 2.3)"
280 input DM
290 rem SF=T=safety factor
300 SF=EXP((3*SM*DM/3.1415927^2)*LOG((1-P)/P))
310 HC=AG/SF
320 PP=P*100
330 print "starting from "N" data of "A$" in "B$" we have:"
340 print "HC"PP" ="HC B$ "     (safety factor, T="SF")"
400 print "          INVERSE USE"
410 print "Calculating % of protected species at conc. C"
420 print "input concentration, in"B$"; 0 to stop the program"
430 input C
440 IF C=0 THEN GOTO 1000
450 SP=100*(1-(1+EXP(3.1415927^2*(AV-LOG(C))/(3*SM*DM)))^(-1))
460 print "the number of protected species with concentration"
470 print  C B$ "is "SP" %"
500 GOTO 420
1000 STOP
```

INDEX

Q

R

S

T

U

V

W

X